UNBOUNDING
THE FUTURE

Also by K. Eric Drexler:

Engines of Creation

UNBOUNDING
THE FUTURE

The Nanotechnology Revolution

K. Eric Drexler and Chris Peterson, with Gayle Pergamit

WILLIAM MORROW AND COMPANY, INC.
NEW YORK

Library of Congress Cataloging-in-Publication Data

Drexler, K. Eric. Unbounding the future: the nanotechnology revolution / by
  K. Eric Drexler and Chris Peterson with Gayle Pergamit.
     p. cm.
  Includes bibliographical references (p.286) and index.
  ISBN 0-688-09124-5
  1. Technology. 2. Biotechnology. I. Peterson, Chris.
  II. Pergamit, Gayle. III. Title.
  T45.D74 1991
  600—dc20   91-6341
                                                   CIP

Printed in the United States of America

First Edition

1 2 3 4 5 6 7 8 9 10

BOOK DESIGN BY ROBIN MALKIN

Foreword

Nanotechnology. The science is good, the engineering is feasible, the paths of approach are many, the consequences are revolutionary-times-revolutionary, and the schedule is: in our lifetimes.

But what?

No one knows but what. That's why a book like this is crucial *before* molecular engineering and the routine transformation of matter arrives. The technology will arrive piecemeal and prominently, but the consequences will arrive at a larger scale and often invisibly.

Perspective from within a bursting revolution is always a problem because the long view is obscured by compelling immediacies and the sudden traffic of people new to the subject, some seizing opportunity, some viewing with alarm. Both optimists and pessimists about new technologies are notorious for their tunnel vision.

The temptation is always to focus on a single point of departure or a single feared or desired goal. Sample point of departure: What if we can make anything out of diamond? Sample feared/desired goal: What if molecular-scale medicine lets people live for centuries?

We're not accustomed to asking, What would a world be like where many such things are occurring? Nor do we ask, What *should* such a world be like?

The first word that comes to mind is *careful*. The second is *carnival*. Nanotechnology breakthroughs are likely to be self-accelerating and self-proliferating, much as information technology advances have been for the past several decades (and will continue to be, especially as nanotech kicks in). We could get a seething texture of constant innovation and surprise, with desired results and unexpected side effects colliding in all directions.

How do you have a careful carnival? *Unbounding the Future* spells out some of the answer.

I've been watching the development of Eric Drexler's ideas since 1975, when he was an MIT undergraduate working on space technologies (space settlements, mass drivers, and solar sailing). Where I was watching from was the "back-to-basics" world of the *Whole Earth Catalog* publications, which I edited at the time. In that enclave of environmentalists and world-savers one of our dirty words was *technofix*. A technofix was deemed always bad because it was a shortcut—an overly focused directing of high tech at a problem with no concern for new and possibly worse problems that the solution might create.

But some technofixes, we began to notice, had the property of changing human perspective in a healthy way. Personal computers empowered individuals and took away centralized control of communication technology. Space satellites—at first rejected by environmentalists—proved to be invaluable environmental surveillance tools, and their images of Earth from space became an engine of the ecology movement.

I think nanotechnology also is a perspective shifter. It is a set of technologies so fundamental as to amount to a whole new domain of back to basics. We must rethink the uses of materials and tools in our lives and civilizations.

Eric showed himself able to think on that scale with his 1986 book *Engines of Creation*. In it he proposed that the potential chaos and hazard of nanotech revolutions required serious anticipatory debate, and for an initial forum he and his wife, Chris Peterson, set up the Foresight Institute. I wrote to Foresight for literature and soon found myself on its board of advisers.

From that vantage point I watched the growing technical challenges to the plausibility of nanotechnology (I also encouraged a few) as people began to take the prospects seriously. The easy challenges were refuted politely. The hard ones changed and improved the body of ideas. None shot it down. Yet.

I also watched the increasing reports from the various technical disciplines of research clearly headed toward nanotech capabilities, mostly by people who had no awareness of one another. I urged Eric and Chris to assemble them at a conference. The First Foresight Conference on Nanotechnology took place in 1989 at Stanford University with a good mix of technical and cultural issues addressed. That convergence quickened the pace of anticipation and research. This book now takes an admirable next step.

As I've learned from the Global Business Network, where I work part-time helping multinational corporations think about their futures, all futurists soon discover that correct prediction is impossible. And forcing the future in a desired direction is also impossible. What does that leave forethought to do? One of the most valuable tools has proved to be what is called scenario planning, in which dramatic, divergent stories of relevant futures are spun out. Divergent strategies to handle them are proposed, and the scenarios and strategies are played against each other until the scenarios are coherent, plausible, surprising, insightful, and checkable against real events as they unfold. "Robust" (adaptable) strategies are supposed to emerge from the process.

This book delivers a rich array of micro-scenarios of nanotechnology at work, some thrilling, some terrifying, all compelling. Probably none represent exactly what will happen, but in aggregate they give a deep sense of the kind of thing that will happen. Strategies of how to stay ahead of the process are proposed, but the ultimate responsibility for the wholesome use and development of nanotechnology falls upon every person aware of it. That now includes you.

—Stewart Brand

Preface

Antibiotics, aircraft, satellites, nuclear weapons, television, mass production, computers, a global petroleum economy—all the familiar revolutions of twentieth-century technology, with their growing consequences for human life and the Earth itself, have emerged within living memory. These revolutions have been enormous, yet the next few decades promise far more. The new prospects aren't as familiar, and can't be: they haven't happened yet. Our aim in this book, though, is to see what we can see, to try to understand not the events of the unknown and unknowable future but distinct, knowable *possibilities* that will shape what the future can become.

Twentieth-century technology is headed for the junk heap, or perhaps the recycling bins. It has changed life; its replacement will change life again, but differently. This book attempts to trace at least a few of the important consequences of the coming revolution in *molecular nanotechnology*, including consequences for the environment, medicine, warfare, industry, society, and life on Earth. We'll paint a picture of the technology itself—its parts, processes, and abilities—but the technology will be a detail in a larger whole.

A short summary of what molecular nanotechnology will mean is *thorough and inexpensive control of the structure of matter.* Pollution, physical disease, and material poverty all stem from poor control of the structure of matter. Strip mines, clear-cutting, refineries, paper mills, and oil wells are some of the crude twentieth-century technologies that will be replaced. Dental drills and toxic chemotherapies are others.

As always, there is both promise of benefit and danger of abuse. As has become routine, the United States is slipping behind by not looking ahead. As never before, foresight is both vital and possible.

I've made the technical case for the feasibility of molecular nanotechnology elsewhere, and this case has been chewed over by scientists and engineers since the mid-1980s. (The technical bibliography outlines some of the relevant literature.) The idea of molecular nanotechnology is now about as well accepted as was the idea of flying to the Moon in the pre–space age year of 1950, nineteen years before the *Apollo 11* landing and seven years before the shock of *Sputnik.* Those who understand it expect it to happen, but without the cost and uncertainty of a grand national commitment.

Our goal in this book is to describe what molecular nanotechnology will mean in practical terms, so that more people can think more realistically about the future. Decisions on how to develop and control powerful new technologies are too important to be left by default to a handful of specialized researchers, or to a hasty political process that flares into action at the last minute when the *Sputnik* goes up. With more widespread understanding and longer deliberation, political decisions are more likely to serve the common good.

I would never have written a book like this on my own; I lean in a more abstract direction. Combined blame and thanks belong to my coauthors, Chris Peterson and Gayle Pergamit, for making this book happen and for clothing the bones of technology in the flesh of human possibilities.

—K. ERIC DREXLER
Stanford University

Authors' Note

Many of the following chapters combine factual descriptions with future scenerios based on those facts. Facts and possibilities by themselves can be dry and disconnected from human affairs; scenarios are widely used by business strategists to link facts and possibilities into coherent, vital pictures. We adopt them for this purpose. *Scenarios are distinguished from the surrounding text by indentation.* Where they speak of technologies, they represent our understanding of what is possible. Where they speak of events occurring before 1991, they represent our understanding of what has already happened. Other elements of scenarios, however, are there to tell a story. The story in the first two paragraphs, set in 1990, is fact.

Contents

Foreword by Stewart Brand   5
Preface   9
Authors' Note   11
1   Looking Forward   17
2   The Molecular World   44
3   Bottom-Up Technology   57
4   Paths, Pioneers, and Progress   84
5   The Threshold of Nanotechnology   117
6   Working with Nanotechnology   127
7   The Spiral of Capability   147
8   Providing the Basics, and More   170
9   Restoring the Environment   181
10 Nanomedicine   199
11 Limits and Downsides   226
12 Safety, Accidents, and Abuse   246
13 Policy and Prospects   265
Afterword: Taking Action   279
Further Reading   286
Technical Bibliography   289
Glossary   293
Acknowledgments   297
Index   299

UNBOUNDING
THE FUTURE

ONE

Looking Forward

The Japanese professor and his American visitor paused in the rain to look at a rising concrete structure on a university campus in the Tokyo suburbs near Higashikoganei Station. "This is for our Nanotechnology Center," Professor Kobayashi said. The professor's guest complimented the work as he wondered to himself, When would an American professor be able to say the same?

This Nanotechnology Center was being built in the spring of 1990, as Eric Drexler was midway through a hectic eight-day trip, giving talks on nanotechnology to researchers and seeing dozens of university and consortium research laboratories. A Japanese research society had sponsored the trip, and the Ministry of International Trade and Industry (MITI) had organized a symposium around the visit—a symposium on molecular machines and nanotechnology. Japanese research was forging ahead, aiming to develop "new modes of science and technology in harmony with nature and human society," a new technology for the twenty-first century.

There is a view of the future that doesn't fit with the view in the newspapers. Think of it as an alternative, a turn in the road of future

17

history that leads to a different world. In that world, cancer follows polio, petroleum follows whale oil, and industrial technology follows chipped flint—all healed or replaced. Old problems vanish, new problems appear: down the road are many alternative worlds, some fit to live in, some not. We aim to survey this road and the alternatives, because to arrive at a world fit to live in, we will all need a better view of the open paths.

How does one begin to describe a process that can replace the industrial system of the world? Physical possibilities, research trends, future technologies, human consequences, political challenges: this is the logical sequence, but none of these makes a satisfactory starting point. The story might begin with research at places like IBM, Du Pont, and the ERATO projects at Tsukuba and RIKEN, but this would begin with molecules, seemingly remote from human concerns. At the core of the story is a kind of technology—"molecular nanotechnology" or "molecular manufacturing"—that appears destined to replace most of technology as we know it today, but it seems best not to begin in the middle. Instead, it seems best to begin with a little of each topic, briefly sketching consequences, technologies, trends, and principles before diving into whole chapters on one aspect or another. This chapter provides those sketches and sets the stage for what follows.

All this can be read as posing a grand "What if?" question: *What if molecular manufacturing and its products replace modern technology?* If they don't, then the question merely invites an entertaining and mind-stretching exercise. But if they do, then working out good answers in advance may tip the balance in making decisions that determine the fate of the world. Later chapters will show why we see molecular manufacturing as being almost inevitable, yet for now it will suffice if enough people give enough thought to the question "What if?"

A SKETCH OF TECHNOLOGIES

Molecular nanotechnology: Thorough, inexpensive control of the structure of matter based on molecule-by-molecule control of products and byproducts; the products and processes of molecular manufacturing.

Technology-as-we-know-it is a product of industry, of manufacturing and chemical engineering. Industry-as-we-know-it takes things from nature—ore from mountains, trees from forests—and coerces them into forms that someone considers useful. Trees become lumber, then houses. Mountains become rubble, then molten iron, then steel, then cars. Sand becomes a purified gas, then silicon, then chips. And so it goes. Each process is crude, based on cutting, stirring, baking, spraying, etching, grinding, and the like.

Trees, though, are not crude: To make wood and leaves, they neither cut, grind, stir, bake, spray, etch, nor grind. Instead, they gather solar energy using molecular electronic devices, the photosynthetic reaction centers of chloroplasts. They use that energy to drive *molecular machines*—active devices with moving parts of precise, molecular structure—which process carbon dioxide and water into oxygen and molecular building blocks. They use other molecular machines to join these molecular building blocks to form roots, trunks, branches, twigs, solar collectors, and more molecular machinery. Every tree makes leaves, and each leaf is more sophisticated than a spacecraft, more finely patterned than the latest chip from Silicon Valley. They do all this without noise, heat, toxic fumes, or human labor, and they consume pollutants as they go. Viewed this way, trees are high technology. Chips and rockets aren't.

Trees give a hint of what molecular nanotechnology will be like, but nanotechnology won't be biotechnology because it won't rely on altering life. Biotechnology is a further state in the domestication of living things. Like selective breeding, it reshapes the genetic heritage of a species to produce varieties more useful to people. Unlike selective breeding, it inserts new genes. Like biotechnology—or ordinary trees—molecular nanotechnology will use molecular machinery, but unlike biotechnology, it will not rely on genetic meddling. It will be not an extension of biotechnology, but an alternative or a replacement.

Molecular nanotechnology could have been conceived and analyzed—though not built—based on scientific knowledge available forty years ago. Even today, as development accelerates, understanding grows slowly because molecular nanotechnology merges fields that have been strangers: the molecular sciences, working at the threshold of the quantum realm, and mechanical engineering, still mired in the grease and crudity of conventional technology. Nanotechnology will be a technology of new molecular machines, of gears and shafts and bearings that move and work with parts shaped in accord with the wave equations at the foundations of natural law. Mechanical engineers don't design molecules. Molecular scientists seldom design machines. Yet a new field will grow—is growing today—in the gap between. That field will replace both chemistry as we know it and mechanical engineering as we know it. And what is manufacturing today, or modern technology itself, but a patchwork of crude chemistry and crude machines?

Chapter 2 will paint a concrete picture of molecular machines and molecular manufacturing, but for now analogy will serve. Picture an automated factory, full of conveyor belts, computers, rollers, stampers, and swinging robot arms. Now imagine something like that factory, but a million times smaller and working a million times faster, with parts and workpieces of molecular size. In this factory, a "pollutant" would be a loose molecule, like a ricocheting bolt or washer, and loose molecules aren't tolerated. In many ways, the factory is utterly unlike a living cell: not fluid, flexible, adaptable, and fertile, but rigid, preprogrammed and specialized. And yet for all of that, this microscopic molecular factory emulates life in its clean, precise molecular construction.

Advanced molecular manufacturing will be able to make almost anything. Unlike crude mechanical and chemical technologies, molecular manufacturing will work from the bottom up, assembling intricate products from the molecular building blocks that underlie everything in the physical world.

Nanotechnology will bring new capabilities, giving us new ways to make things, heal our bodies, and care for the environment. It will also bring unwelcome advances in weaponry and give us yet more ways to foul up the world on an enormous scale. It won't automatically solve our problems: even powerful technologies merely give us more power. As usual, we have a lot of work ahead of us, and a lot of hard decisions to make if we hope to harness new developments to good ends. The main reason to pay attention to nanotechnology now, before it exists, is to get a head start on understanding it and what to do about it.

A SKETCH OF CONSEQUENCES

The United States has become famous for its obsession with the next year's elections and the next quarter's profits, and the future be damned. Nonetheless, we are writing for normal human beings who feel that the future matters—ten, twenty, perhaps even thirty years from now—for people who care enough to try to shift the odds for the better. Making wise choices with an eye to the future requires a realistic picture of what the future can hold. What if most pictures of the future today are based on the wrong assumptions?

Here are a few of today's common assumptions, some so familiar that they are seldom stated:

- Industrial development is the only alternative to poverty.
- Many people must work in factories.
- Greater wealth means greater resource consumption.
- Logging, mining, and fossil-fuel burning must continue.
- Manufacturing means pollution.
- Third World development would doom the environment.

These all depend on a more basic assumption:

- *Industry as we know it cannot be replaced.*

Some further common assumptions:

- The twenty-first century will basically bring more of the same.
- Today's economic trends will define tomorrow's problems.
- Spaceflight will never be affordable for most people.
- Forests will never grow beyond Earth.
- More advanced medicine will always be more expensive.
- Even highly advanced medicine won't be able to keep people healthy.
- Solar energy will never become really inexpensive.
- Toxic wastes will never be gathered and eliminated.
- Developed land will never be returned to wilderness.
- There will never be weapons worse than nuclear missiles.
- Pollution and resource depletion will eventually bring war or collapse.

These, too, depend on a more basic assumption:

- *Technology as we know it will never be replaced.*

These commonplace assumptions paint a future full of terrible dilemmas, and the notion that a technological change will let us escape from them smacks of the idea that some technological fix can save the industrial system. The prospect, though, is quite different: The industrial system won't be fixed, it will be junked and recycled. The prospect isn't more industrial wealth ripped from the flesh of the Earth, but green wealth unfolding from processes as clean as a growing tree. Today, our industrial technologies force us to choose better quality *or* lower cost *or* greater safety *or* a cleaner environment. Molecular manufacturing, however, can be used to improve quality *and* lower costs *and* increase safety *and* clean the environment. The coming revolutions in technology will transcend many of the old, familiar dilemmas. And yes, they will bring fresh, equally terrible dilemmas.

Molecular nanotechnology will bring thorough and inexpensive control of the structure of matter. We need to understand molecular nanotechnology in order to understand the future capabilities of the human race. This will help us see the challenges ahead, and help us plan how best to conserve values, traditions, and ecosystems through effective policies and institutions. Likewise, it can help us see what today's events mean, including business opportunities and possibilities for action. We need a vision of where technology is leading because technology is a part of what human beings are, and will affect what we and our societies can become.

The consequences of the coming revolutions will depend on human actions. As always, new abilities will create new possibilities both for good and for ill. We will discuss both, focusing on how political and economic pressures can best be harnessed to achieve good ends. Our answers will not be satisfactory, but they are at least a beginning.

A SKETCH OF TRENDS

Technology has been moving toward greater control of the structure of matter for millennia. For decades, microtechnology has been building ever-smaller devices, working toward the molecular size scale from the top down. For a century or more, chemistry has been building ever-larger molecules, working up toward molecules large enough to serve as machines. The research is global, and the competition is heating up.

Since the concept of molecular nanotechnology was first laid out, scientists have developed more powerful capabilities in chemistry and molecular manipulation (see Chapter 4). There is now a better picture of how those capabilities can come together in the next steps (see Chapter 5), and of how advanced molecular manufacturing can work (see Chapter 6). Nanotechnology has arrived as an idea and as a research direction, though not yet as a reality.

Naturally occurring molecular machines exist already. Researchers are learning to design new ones. The trend is clear, and it will accelerate because better molecular machines can help build even better molecular machines. By the standards of daily life, the development of molecular nanotechnology will be gradual, spanning years or decades, yet by the ponderous standards of human history it will happen in an eyeblink. In retrospect, the wholesale replacement of twentieth-century technologies will surely be seen as a technological revolution, as a process encompassing a great breakthrough.

Today, we live in the end of the pre-breakthrough era, with pre-breakthrough technologies, hopes, fears, and preoccupations that often seem permanent, as did the Cold War. Yet it seems that the breakthrough era is not a matter for some future generation, but for our own. These developments are taking shape right now, and it would be rash to assume that their consequences will be many years delayed.

In later chapters, we'll say more about what researchers are doing today, about where their work is leading, and about the problems and choices ahead. To get a sense of the consequences, though, requires a picture of what nanotechnology can do. This can be hard to grasp because past advanced technologies—microwave tubes, lasers, superconductors, satellites, robots, and the like—have come trickling out of factories, at first with high price tags and narrow applications. Molecular manufacturing, though, will be more like computers: a flexible technology with a huge range of applications. And molecular manufacturing won't come trickling out of conventional factories as computers did: it will *replace* factories and replace or upgrade their products. This is something new and basic, not just another twentieth-century gadget. It will arise out of twentieth-century trends in science, but it will break the trend-lines in technology, economics, and environmental affairs.

Calculators were once thousand-dollar desktop clunkers, but microelectronics made them fast and efficient, sized to a child's pocket and priced to a child's budget. Now imagine a revolution of similar magnitude, but applied to everything else.

MORE CONSEQUENCES: SCENES FROM A POST-BREAKTHROUGH WORLD

What nanotechnology will mean for human life is beyond our predicting, but a good way to understand what it *could* mean is to paint scenarios. A good scenario brings together different aspects of the world (technologies, environments, human concerns) into a coherent whole. Major corporations use scenarios to help envision the paths that the future may take—not as forecasts, but as tools for thinking. In playing the "What if?" game, scenarios present trial answers and pose new questions.

The following scenarios can't represent what will happen, because no one knows. They can, however, show how post-breakthrough capabilities could mesh with human life and Earth's environment. The results will likely seem quaintly conservative from a future perspective, however much they seem like science fiction today. The issues behind these scenarios will be discussed in later chapters.

SCENARIO: SOLAR ENERGY

> In Fairbanks, Alaska, Linda Hoover yawns and flips a switch on a dark winter morning. The light comes on, powered by stored solar electricity. The Alaska oil pipeline shut down years ago, and tanker traffic is gone for good.

Nanotechnology can make solar cells efficient, as cheap as newspaper, and as tough as asphalt—tough enough to use for resurfacing roads, collecting energy without displacing any more grass and trees. Together with efficient, inexpensive storage cells, this will yield low-cost power (but no, not "too cheap to meter"). Chapter 9 discusses prospects for energy and the environment in more depth.

Scenario: Medicine that Cures

Sue Miller of Lincoln, Nebraska, has been a bit hoarse for weeks, and just came down with a horrid head cold. For the past six months, she's been seeing ads for At Last!®: The Cure for the Common Cold, so she spends her five dollars and takes the nose-spray and throat-spray doses. Within three hours, 99 percent of the viruses in her nose and throat are gone, and the rest are on the run. Within six hours, the medical mechanisms have become inactive, like a pinch of inhaled but biodegradable dust, soon cleared from the body. She feels much better and won't infect her friends at dinner.

The human immune system is an intricate molecular mechanism, patrolling the body for viruses and other invaders, recognizing them by their foreign molecular coats. The immune system, though, is slow to recognize something new. For her five dollars, Sue bought 10 billion molecular mechanisms primed to recognize not just the viruses she had already encountered, but each of the five hundred most common viruses that cause colds, influenza, and the like.

Weeks have passed, but the hoarseness Sue had before her cold still hasn't gone away; it gets worse. She ignores it through a long vacation, but once she's back and caught up, Sue finally goes to see her doctor. He looks down her throat and says, "Hmmm." He asks her to inhale an aerosol, cough, spit in a cup, and go read a magazine. The diagnosis pops up on a screen five minutes after he pours the sample into his cell analyzer. Despite his knowledge, his training and tools, he feels chilled to read the diagnosis: a malignant cancer of the throat, the same disease that has cropped up all too often in his own mother's family.

He touches the "Proceed" button. In twenty minutes, he looks at the screen to check progress. Yes, Sue's cancerous cells are all of one basic kind, displaying one of the 16,314 known molecular markers for malignancy. They can be recognized, and since they can be recognized, they can be destroyed by standard molecular machines primed to react to those markers. The doctor instructs the cell analyzer to prime some "immune machines" to go after the cancer cells. He tests them on cells from the sample, watches, and sees that they work as expected, so he has the analyzer prime up some more.

Sue puts the magazine down and looks up. "Well, Doc, what's the word?" she asks.

"I found some suspicious cells, but this should clear it up," he says. He gives her a throat spray and an injection. "I'd like you to come back in three weeks, just to be sure."

"Do I have to?" she asks.

"You know," he lectures her, "we need to make sure it's gone. You really shouldn't let things like this go so far before coming in."

"Yes, fine, I'll make the appointment," she says. Leaving the office, Sue thinks fondly of how old-fashioned and conservative Dr. Fujima is.

The molecular mechanisms of the immune system already destroy most potential cancers before they grow large enough to detect. With nanotechnology, we will build molecular mechanisms to destroy those that the immune system misses. Chapter 10 discusses medical nanotechnologies in more depth.

SCENARIO: CLEANSING THE SOIL

California Scout Troop 9731 has hiked for six days, deep in the second-wilderness forests of the Pacific Northwest.

"I bet we're the first people ever to walk here," says one of the youngest scouts.

"Well, maybe you're right about *walking*," says Scoutmaster Jackson, "but look up ahead—what do you see, scouts?"

Twenty paces ahead runs a strip of younger trees, stretching left and right until it vanishes among the trunks of the surrounding forest.

"Hey, guys! Another old logging road!" shouts an older scout. Several scouts pull probes from their pockets and fit them to the ends of their walking sticks. Jackson smiles: It's been ten years since a California troop found anything this way, but the kids keep trying.

The scouts fan out, angling their path along the scar of the old road, poking at the ground and watching the readouts on the stick handles. Suddenly, unexpectedly, comes a call: "I've got a signal! Wow—I've got PCBs!"

In a moment, grinning scouts are mapping and tracing the spill. Decades ago, a truck with a leaking load of chemical waste snuck down the old logging road, leaving a thin toxic trail. That trail leads them to a deep ravine, some rusted drums, and a nice wide path of invisible filth. The excitement is electrifying.

Setting aside their maps and orienteering practice, they unseal a satellite locator to log the exact latitude and longitude of the site, then send a message that registers their cleanup claim on the ravine. The survey done, they head off again, eagerly planning a return trip to earn the now-rare Toxic Waste Cleanup Merit Badge.

Today, tree farms are replacing wilderness. Tomorrow, the slow return to wilderness may begin, when nature need no longer be seen as a storehouse of natural resources to be plundered. Chapter 9 will discuss just how little need be taken from nature to provide humans with wealth, and how post-breakthrough technologies can remove from nature the toxic residues of twentieth-century mistakes.

SCENARIO: POCKET SUPERCOMPUTERS

At the University of Michigan, Joel Gregory grabs a molecular rod with both hands and twists. It feels a bit weak, and a ripple of red reveals too much stress in a strained molecular bond halfway down its length. He adds two atoms and twists the rod again: all greens and blues, much better. Joel plugs the rod into the mechanical arm he's designing, turns up the temperature, and sets the whole thing in motion. A million atoms dance in thermal vibration, gears spin, and the arm swings to and fro in programmed motion. It looks good. A few parts are still mock-ups, but doing a thesis takes time, and he'll work out the rest of the molecular details later. Joel strips off the computer display goggles and gloves and blinks at the real world. It's time for a sandwich and a cup of coffee. He grabs the computer itself, stuffs it into his pocket, and heads for the student center.

Researchers already use computers to build models of molecules, and "virtual reality systems" have begun to appear, enabling a user to walk around the image of a molecule and "touch" it, using computer-controlled gloves and goggles. We can't build a supercomputer able to model a million-atom machine yet—much less build a pocket supercomputer—but computers keep shrinking in size and cost. With nanotechnology to make molecular parts, a computer like Joel's will become easy to build. Today's supercomputers will seem like hand-cranked adding machines by comparison. Chapters 2 and 3 take a closer look at a simulated molecular world.

SCENARIO: GLOBAL WEALTH

Behind a village school in the forest a stone's throw from the Congo River, a desktop computer with a thousand times the power of an early 1990s supercomputer lies half-buried in a recycling bin. Indoors, Joseph Adoula and his friends have finished their day's studies; now they are playing together in a vivid game universe using personal computers each a million times more powerful than the clunker in the trash. They stay late in air-conditioned comfort.

Trees use air, soil, and sunlight to make wood, and wood is cheap enough to burn. Nanotechnology can do likewise, making products as cheap as wood—even products like supercomputers, air conditioners, and solar cells to power them. The resulting economics may even keep tropical forests from being burned. Chapter 7 will discuss how costs can fall low enough to make material wealth for the Third World easy to achieve.

SCENARIO: CLEANSING THE AIR

In Earth's atmosphere, the twentieth-century rise in carbon-dioxide levels has halted and reversed. Fossil fuels are obsolete, so pollution rates have lessened. Efficient agriculture has freed fertile land for reforestation, so growing trees are cleansing the atmosphere. Surplus solar power from the world's repaved roads is being used to break down excess carbon dioxide at a rate of 5 billion tons per year. Climates are returning to normal, the seas are receding to their historical shores, and ecosystems are beginning the slow process of recovery. In another twenty years, the atmosphere will be back to the pre-industrial composition it had in the year 1800.

Chapter 9 will discuss environmental cleanup, from reducing the sources to cleaning up the messes already in place.

SCENARIO: TRANSPORTATION OUTWARD

Jim Salin's afternoon flight from Dulles International is on the ground, late for departure. Impatiently, Jim checks the time: any later, and he'll miss his connecting flight.

At last, the glassy-surfaced craft rolls down the runway. With gliderlike wings, it lifts its fat body and climbs steeply toward the east. A few pages into his novel, Jim is interrupted by a second recitation of safety instructions and the captain's announcement that they'll try to make up for lost time. Jim settles back in his seat as the main engines kick in, the wings retract, the acceleration builds, and the sky darkens to black. Like the highest-performance rockets of the 1980s, Jim's liner produces an exhaust of pure water vapor. Spaceflight has become clean, safe, and routine. And every year, more people go up than come down.

The cost of spaceflight is mostly the cost of high-performance, reliable hardware. Molecular manufacturing will make aerospace structures from nearly flawless, superstrong materials at low cost. Add inexpensive fuel, and space will become more accessible than the other side of the ocean is today. Chapter 8 discusses the prospects for opening the world beyond Earth.

SCENARIO: RESTORING SPECIES

Restoration Day Ceremonies are always moving events. For some reason, the old people always cry, even though they say they're happy.

Crying, Tracy Stiegler thinks, *doesn't make any sense.* She looks again through the camouflage screen over the sandy Triangle Keys beach, gazing across the Caribbean toward the Yucatán Peninsula. *Soon this will be theirs again, and that's all to the good.*

Tracy and the other scientists from BioArchive have positions of honor in today's Restoration Day Ceremony. Since the mid-twentieth century, there had been no living Caribbean monk seals, only grisly relics of the years of their slaughter: seal furs and dry museum specimens. Tracy's team struggled for years, gathering these relics and studying them with molecular instruments. It had been known for decades—since the 1980s—that genes are tough enough to survive in dried skin, bone, horn, and eggshell. Tracy's team had collected genes and rebuilt cells.

They worked for years, and gave thanks to the strict protection—late, but good enough—that saved one related species. At last, a Hawaiian monk seal had given birth to a genetically pure Caribbean monk seal, twin to a seal long dead. And now there were five hundred, some young, some middle-aged, with decent genetic diversity and five years' experience living in the confines of a coastal ecological station.

Today, with raucous voices, they are moving out into the world to reclaim their ecological niche. As Tracy watches, she thinks of the voices that will never be heard again: of the species, known and unknown, that left not even a bloody scrap to be cherished and restored. Thousands (millions?) of species had simply been brushed into extinction as habitats were destroyed by farming and logging. People knew—for *years* they had known—that freezing or drying would save genes. And they knew of the ecological destruction, and they knew they weren't stopping it. And the ignorant bastards didn't even keep samples.

Tracy discovers that she, too, cries at Restoration Day Ceremonies.

People will surely push biomedical applications of nanotechnology far and fast for human health care. With a bit more pushing, this technology base will be good enough to restore some species now thought lost forever, to repair some of the damage human beings have done to the web of life. It would be better to preserve ecosystems and species intact, but restoration, even of a few species, will be far better than nothing. Some samples from endangered species *are* being kept today, but not enough, and mostly for the wrong reasons. Chapter 9 will take a closer look at ecosystem restoration, and what future prospects mean for action taken today.

SCENARIO: AN UNSTABLE ARMS RACE

Disputes over technology development and trade had soured relationships between Singapore and the Japan-United States alliance. Diplomatic inquiries regarding peculiar seismic and sonar readings in the South China Sea had just begun when they suddenly became irrelevant: an estimated one billion tons of unfamiliar, highly automated military hardware appeared in coastal waters around the world. Accusations began to fly between Congress and PeaceWatch personnel: "If you'd done your jobs—" "If you'd let us do our jobs—"

And so, in late February, Singapore emerged as a military superpower.

Low cost, high-quality, high-speed production can be applied to many purposes, not all attractive. Nanotechnology has enormous potential for abuse.

TECHNOLOGIES REVISITED

Molecules matter because matter is made of molecules, and everything from air to flesh to spacecraft is made of matter. When we learn how to arrange molecules in new ways, we can make new things, and make old things in new ways. Perhaps this is why Japan's MITI has identified "control technologies for the precision arrangement of molecules" as a basic industrial technology for the twenty-first century. Molecular nanotechnology will give thorough control of matter on a large scale at low cost, shattering a whole set of technological and economic barriers more or less at one stroke.

A molecule is an object consisting of a collection of atoms held together by strong bonds (one-atom molecules are a special case). "Molecule" usually refers to an object with a number of atoms small enough to be counted (a few to a few thousand), but strictly speaking a truck tire (for instance) is mostly one big molecule, containing something like 1,000,000,000,000,000,000,000,000,000 atoms. Counting this many atoms aloud would take about 10,000,000,000 *billion* years.

Scientists and engineers still have no direct, convenient way to control molecules, basically because human hands are about 10 million times too large. Today, chemists and materials scientists make molecular structures indirectly, by mixing, heating, and the like. The idea of nanotechnology begins with the idea of a *molecular assembler*, a device resembling an industrial robot arm but built on a microscopic scale. A general-purpose molecular assembler will be a jointed mechanism built from rigid molecular parts, driven by motors, controlled by computers, and able to grasp and apply molecular-scale tools. Molecular assemblers can be used to build other molecular machines—they can even build more molecular assemblers. Assemblers and other machines in molecular-manufacturing systems will be able to make almost anything, if given the right raw materials. In effect, molecular assemblers will provide the microscopic "hands" that we lack today. (Chemists are asked to forgive this literary license; the specific details of molecular binding and bonding don't change the conclusion.)

Nanotechnology will give better control of molecular building blocks, of how they move and go together to form more complex objects. Molecular manufacturing will make things by building from the bottom up, starting with the smallest possible building blocks. The *nano* in nanotechnology comes from *nanos*, the Greek word for dwarf. In science, the prefix *nano-* means one-billionth of something, as in nanometer and nanosecond, which are typical units of size and time in the world of molecular manufacturing. When you see it tacked onto the name of an object, it means that the object is made by patterning matter with molecular control: nanomachine, nanomotor, nanocomputer. These are the smallest, most precise devices that make sense based on today's science.

(Be cautious of other usages, though—some researchers have begun to use the *nano-* prefix to refer to other small-scale technologies in the laboratory today. In this book, *nanotechnology* means the *precise, molecular* nanotechnology of the future. British usage also applies the term to the small-scale and high-precision technologies of today—even to precision grinding and measurement. The latter are useful, but hardly revolutionary.)

Digital electronics brought an information-processing revolution by handling information quickly and controllably in perfect, discrete pieces: bits and bytes. Likewise, nanotechnology will bring a matter-processing revolution by handling matter quickly and controllably in perfect, discrete pieces: atoms and molecules. The digital revolution has centered on a device able to make any desired pattern of bits: the programmable computer. Likewise, the nanotechnological revolution will center on a device able to make (almost) any desired pattern of atoms: the programmable assembler. The technologies that plague us today suffer from the messiness and wear of an old phonograph record. Nanotechnology, in contrast, will bring the crisp, digital perfection of a compact disc.

A ROAD MAP

The next two sections say a bit more about why nanotechnology is already worth your attention and about whether it's possible to understand anything about the future. Later chapters answer questions like the following:

- Who is working on nanotechnology? What are they doing, and why?
- How can this work come together to provide breakthrough capabilities? When might this happen? What developments should we watch for?
- How will nanotechnology work? Who will be able to use it?
- What will it mean for the economy? For medicine? For the environment?
- What are its risks? What basic regulations will we need? What will it mean for the global arms race?
- What might go wrong as this technology emerges, and what can we do about it?

In a democratic society, only a few people need an in-depth understanding of how a technology works, but many people need to understand what it can do. In the next chapter, we'll lead off by describing the molecular world and how it works—after all, everything around us and inside us is made of molecules—but the main story is about what this technology will mean for human beings and the biosphere.

WHY TALK ABOUT IT?

It is these concerns—the implications of nanotechnology for our lives, the environment, and the future—that guided the writing of this book.

Nanotechnology can bring great achievements and solve great problems, but it will likewise present opportunities for enormous abuse. Research progress is necessary, but so is an informed and cautious public.

Our motivation in presenting these ideas is as much a fear of potential harm, and a wish to avoid it, as a longing for the potential good and a wish to seek it. Even so, we will dwell on the good that nanotechnology can bring and give only an outline of the obvious potential harm. The coming revolution can best be managed by people who share not only a picture of what they wish to avoid, but of what they can achieve. If we as a society have a clear view of a route to follow, we won't need a precise catalog of every cliff and mine field to the side of the road.

Some will hear this emphasis and call us optimistic. But would it really be wise to dwell on exactly how a technology can be abused? Or to draw up blueprints, perhaps?

Still, sitting here, preparing to tell this story, is an uncomfortable place for a researcher to be. In his book *How Superstition Won and Science Lost*, historian John C. Burnham tells of the century-long retreat of scientists from what they once saw as their responsibility: presenting the content and methods of science to a broad audience, for the public good. Today, the culture of science takes a dim view of "popularization." If you can write in plain English, this is taken as evidence that you can't do math, and vice versa. Robert Pool, a member of the news staff of the most prestigious American scientific journal, *Science*, acknowledges this negative attitude in writing that "some researchers, either by choice or just by being in the wrong place at the wrong time, make it into the public eye." So how can a researcher keep out of trouble? If you stumble on something important, wrap it in jargon. If people realize that it's important, run and hide. Robert Pool gently urges scientists to become more involved, but the social pressures in the research community are heavily in the other direction.

In response to this negative attitude toward "popularization," we can only ask that scientists and engineers try to act in a thoroughly professional fashion when judging a given proposal—which is to say, that they pay scrupulous attention to the scientific and technical facts. This means judging the validity of technical ideas based on their factual merits, and not on their (occasionally readable) style of presentation, or on the emotional response they may stir up. Nanotechnology matters to people, and they deserve to know about its flesh-and-blood human consequences, its impact on society and nature. We urge scientifically inclined readers to consult the Technical Bibliography at the end of the book, and then to point out any major errors they can find in the technical papers on this topic. We urge nonscientists who encounter scientifically knowledgeable critics to ask for *specific, technical* criticisms. We'll discuss some of the criticisms made to date in Chapter 3. Years of discussion with scientists and engineers—in public, in private, at conferences, and through the press—indicate that the case for nanotechnology is solid. Japanese and European industry, government, and academic researchers are forging ahead on the road to nanotechnology, and more and more U.S. research is applicable. Some researchers have even begun to call it an obvious goal.

WORDS THAT BLOCK THINKING

Americans, so often in the forefront of science and technology, have a curious difficulty in thinking about the future. Language seems to have something to do with it.

If something sounds futurelike, we call it "futuristic." If that doesn't stop the conversation, we say that it "sounds like science fiction." These descriptions remind listeners of laughable 1950s fantasies like rockets to the Moon, video telephones, ray guns, robots, and the like. Of course, all these became real in the 1960s, because the science *wasn't* fiction. Today, we can see not only how to build additional science-fictional devices, but—more important, for better or worse—how to make them cheap and abundant. We need to think about the future, and name-calling won't help.

Curiously, the Japanese language seems to lack a disparaging word

A serious problem. (Calvin and Hobbes. Copyright © 1989 by Universal Press Syndicate. Reprinted with permission. All rights reserved.)

for "futurelike." Ideas for future technologies may be termed *mirai no* ("of the future," a hope or a goal), *shōrai-teki* (an expected development, which might be twenty years away), or *kūsō no* ("imaginary" only, because contrary to physical law or economics). To think about the future, we need to distinguish *mirai no* and *shōrai-teki*, like nanotechnology, from mere *kūsō no* like antigravity boots.

A final objection is the claim that there's no point in trying to think about the future, because it is all too complex and unpredictable. This is too sweeping, but has more than a little truth. It deserves a considerable response.

THE DIFFICULTY OF LOOKING FORWARD

If our future will include nanotechnology, then it would be useful to understand what it can do, so that we can make more sensible plans for our families, careers, companies, and society. But many intelligent people will respond that understanding is impossible, that the future is just too unpredictable. This depends, of course, on what you're trying to predict:

The weather a month from now? Forget it; weather is too chaotic.

The position of the Moon a century from now? Easy; the Moon's orbit is like clockwork.

Which personal-computer company will lead twenty years from now? Good luck; major companies today didn't even exist twenty years ago.

That personal computers will become enormously more powerful? A virtual certainty.

And so on. If you aim to say something sensible about the future of technology, the trick is to ask the right questions and to avoid the standard pitfalls. In his book *Megamistakes: Forecasting and the Myth of Rapid Technological Change*, Steven Schnaars surveys these pitfalls and their effects on past predictions. Borrowing and adapting some of his generalizations, here are our suggestions for how to blunder into a Megamistake in forecasting:

- Ignore the scientific facts, or guess.
- Forget to ask whether anyone wants the projected product or situation.
- Ignore the costs.
- Try to predict which company or technology will win.

In looking at what to expect from nanotechnology—or any technology—all of these must be avoided, since they can lead to some grand absurdities. In a classic demonstration of the first error, someone once concocted the notion that pills would someday replace food. But people need energy to live, and energy means calories, which means fuel, which takes up room. To subsist on pills, you'd need to gobble them by the fistful. This would be like eating a tasteless kibbled dog food, which was hardly the idea. In short, the pills-for-food prediction ignored the scientific facts. In a similar vein, we once heard promises of a cure for cancer—but this was based on a guess about scientific facts, a guess that "cancer" was in some sense a single disease, which might have a single point of vulnerability and a single cure. This guess was wrong, and progress against cancer has been slow.

Earlier, we presented a scenario that includes the routine cure of a cancer using nanotechnology. This scenario takes account of the currently known facts: Cancers differ, but each kind can be recognized by its molecular markers. Molecular machines can recognize molecular markers, and so can be primed to recognize and destroy specific kinds of cancer cells as they turn up. We will explore medical applications of nanotechnology further in Chapter 10.

Even nanotechnology can't cram a meal into a pill, but this is just as well. The pills-for-food proposal didn't just ignore the facts, it also ignored what people *want*—things like dinner conservation and novel ethnic cuisines. Magazines once promised cities beneath the sea, but who wants to live in the ultimate damp, chilly climate? California and the Sunbelt have somehow proved more popular. And again, we were promised talking cars, but after giving them a try, people prefer luxury cars from companies that promise silence.

Many human wants are easy to predict, because they are old and stable: People want better medical care, housing, consumer goods, transportation, education, and so forth, preferably at lower costs, with greater safety, in a cleaner environment. When our limited abilities force us to choose better quality *or* lower cost *or* greater safety *or* a cleaner environment, decisions become sticky. Molecular manufacturing will allow a big step in the direction of better quality *and* lower costs *and* increased safety *and* a cleaner environment. (Choices of *how much of each* will remain.) There is no existing market demand for "nanotechnology," as such, but a great demand for what it can do.

Neglecting costs has also been popular among prognosticators: Building cities under the sea would be expensive, with few benefits. Building in space has more benefits, but would be far more expensive, using past or present technologies. Many bold projections gather dust on shelves because development or manufacturing costs are too high. Some examples include personal robots, flying cars, and Moon colonies—they still sound more like 1950s science fiction than practical possibilities, and cost is one major reason.

Molecular manufacturing is, in part, about cost reduction. As mentioned above, molecular machines in nature make things cheaply, like wood, potatoes, and hay. Trees are more complex than spacecraft, so why should spacecraft stay more expensive? Gordon Tullock, professor of economics and political science at the University of Arizona, says of molecular nanotechnology, "Its economic effect is that we will all be much richer." The prospect of building sophisticated products for the price of potatoes gives reason to pull a lot of old projections down from the shelf. We hope you won't mind the dust when we brush them off for a fresh look.

Even staying within the bounds of known science, focusing on things people want, and paying attention to costs, it's still hard to pick a specific winner. Technology development is like a horse race: everyone knows that some horse will win, but knowing *which* is harder (and worth big bucks). Both corporate managers betting money and researchers betting their careers have to play this game, and they often lose. A technology may work, provide something useful, and be less expensive than last year's alternative, yet still be clobbered in the market by something unexpected but better. To know which technologies will win, you'd have to know *all* the alternatives, whether they've been invented yet or not. Good luck!

We won't try to play that game here. "Nanotechnology" (like "modern industry") describes a huge range of technologies. Nonetheless, nanotechnology in one form or another is a monumentally obvious idea: it will be the culmination of an age-old trend toward more thorough control of the structure of matter. Predicting that some form of nanotechnology will win most technology races is like predicting that some horse will win a horse race (as opposed to, say, a dachshund). A technology based on thorough control of the structure of matter will almost always beat one based on crude control of the structure of matter. Other technologies have already won races in the literal sense of being *first*. Few, however, will win in the sense of being *best*.

EXPLORATORY ENGINEERING

Studies of nanotechnology are today in the *exploratory engineering* phase, and just beginning to move into engineering development. The basic idea of exploratory engineering is simple: combine engineering principles with known scientific facts to form a picture of future technological possibilities. Exploratory engineering looks at future possibilities to help guide our attention in the present. Science—

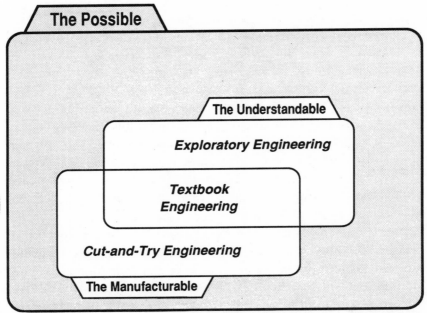

EXPLORATORY ENGINEERING VENN DIAGRAM K. ERIC DREXLER
The outer tagged rectangle represents the set of all technologies permitted by the laws of nature, whether they exist or not, whether they have been imagined or not. Within this set are those technologies that are manufacturable with today's technology, and those that are understandable with today's science. Textbooks teach what is understandable (hence teachable) and manufacturable (hence immediately practical). Practical engineers achieve many successes by cut-and-try methods and put them into production. Exploratory engineers study what will become practical as manufacturing abilities expand to embrace more of the possible.

especially molecular science—has moved fast in recent decades. There is no need to wait for more scientific breakthroughs in order to make engineering breakthroughs in nanotechnology.

The above illustration shows how exploratory engineering relates to more familiar kinds of engineering. Each works within the limits of the possible, which are set by the known and unknown laws of nature. The most familiar kind is the engineering taught in schools: this "textbook engineering" covers technologies that can be both *understood* (so they can be taught) and *manufactured* (so they can be used). Bridge-building and gearbox design fall in this category. Other technologies, however, can be manufactured but aren't understood—any engineer can give examples of things that work when similar things don't, and for no obvious reason. But as long as they *do* work, and work consistently, they can be used with confidence. This is the world of "cut-and-try engineering," so important to modern industry. Bearing lubrication, adhesives, and many manufacturing technologies advance by cut-and-try methods.

Exploratory engineering covers technologies that can be understood but *not* manufactured—yet. Technologies in this category are also familiar to engineers, although normally they design such things only for fun. So much is known about mechanics, thermodynamics, electronics, and so forth that engineers can often calculate what something will do, just from a description of it. Yet there is no reason why everything that can be correctly described must be manufacturable—the constraints are different. Exploratory engineering is as simple as textbook engineering, but neither military planners nor corporate executives see much profit in it, so it hasn't received much attention.

The concepts of molecular manufacturing and molecular nanotechnology are straightforward results of exploratory-engineering research applied to molecular systems. As we observed above, the basic ideas could have been worked out forty years ago, if anyone had bothered. Naturally enough, both scientists and engineers were preoccupied with more immediate concerns. But now, with the threshold of nanotechnology approaching, attention is beginning to focus on where the next steps lead.

Nanotechnology seems to be where the world is headed if technology keeps advancing, and competition practically guarantees that advances will continue. It will open both a huge range of opportunities for benefit and a huge range of opportunities for misuse. We will paint scenarios to give a sense of the prospects and possibilities, but we don't offer predictions of what will happen. Actual human choices and blunders will depend on a range of factors and alternatives beyond what we can hope to anticipate.

TWO

The Molecular World

N anotechnology will be a bottom-up technology, building upward from the molecular scale. It will bring a revolution in
human abilities like that brought by agriculture or power machinery.
It can even be used to reverse many of the changes brought by agriculture and power machinery. But we humans are huge creatures
with no direct experience of the molecular world, and this can make
nanotechnology hard to visualize, hence hard to understand.

Scientists working with molecules face this problem today. They
can often calculate how molecules will behave, but to *understand*
that behavior, they need more than heaps of numbers: they need
pictures, movies, and interactive simulations, and so they are producing them at an ever-increasing pace. The U.S. National Science
Foundation has launched a program in "scientific visualization," in
part to harness supercomputers to the problem of picturing the molecular world.

Molecules are objects that exert forces on one another. If your
hands were small enough, you could grab them, squeeze them, and
bash them together. Understanding the molecular world is much like

understanding any other physical world: it is a matter of understanding size, shape, strength, force, motion, and the like—a matter of understanding the differences between sand, water, and rock, or between steel and soap bubbles. Today's visualization tools give a taste of what will become possible with tomorrow's faster computers and better "virtual realities," simulated environments that let you tour a world that "exists" only as a model inside the computer. Before discussing nanotechnology and how it relates to the technologies of today, let's try to get a more concrete understanding of the molecular world by describing a simulation embedded in a scenario. In this scenario, events and technologies described as dating from 1990 or before are historically accurate; those with later dates are either projections or mere scenario elements. The descriptive details in the simulation are written to fit designs and calculations based on standard scientific data, so the science isn't fiction.

EXPLORING THE MOLECULAR WORLD

In a scenario in the last chapter, we saw Joel Gregory manipulating molecules in the virtual reality of a simulated world using video goggles, tactile gloves, and a supercomputer. The early twenty-first century should be able to do even better. Imagine, then, that today you were to take a really long nap, oversleep, and wake up decades later in a nanotechnological world.

> In the twenty-first century, even more than in the twentieth, it's easy to make things work without understanding them, but to a newcomer much of the technology seems like magic, which is dissatisfying. After a few days, you want to understand what nanotechnology is, on a gut level. Back in the late twentieth century, most teaching used dry words and simple pictures, but now—for a topic like this—it's easier to explore a simulated world. And so you decide to explore a simulation of the molecular world.
>
> Looking through the brochure, you read many tedious facts

about the simulation: how accurate it is in describing sizes, forces, motions, and the like; how similar it is to working tools used by both engineering students and professionals; how you can buy one for your very own home, and so forth. It explains how you can tour the human body, see state-of-the-art nanotechnology in action, climb a bacterium, etc. For starters, you decide to take an introductory tour: simulations of real twentieth-century objects alongside quaint twentieth-century concepts of nanotechnology.

After paying a small fee and memorizing a few key phrases (any variation of "Get me out of here!" will do the most important job), you pull on a powersuit, pocket a Talking Tourguide, step into the simulation chamber, and strap the video goggles over your eyes. Looking through the goggles, you seem to be in a room with a table you know isn't really there and walls that seem too far away to fit in the simulation chamber. But trickery with a treadmill floor makes the walk to the walls seem far enough, and when you walk back and thump the table, it feels solid because the powersuit stops your hand sharply at just the right place. You can even feel the texture of the carvings on the table leg, because the suit's gloves press against your fingertips in the right patterns as you move. The simulation isn't perfect, but it's easy to ignore the defects. On the table is (or seems to be) an old 1990s silicon computer chip. When you pick it up, as the beginners' instructions suggest, it looks like Figure 1A. Then you say, "Shrink me!" and the world seems to expand.

VISION AND MOTION

You feel as though you're falling toward the chip's surface, shrinking rapidly. In a moment, it looks roughly like Figure 1B, with your thumb still there holding it. The world grows blurrier, then everything seems to go wrong as you approach the molecular level. First, your vision blurs to uselessness—there is light, but it becomes a featureless fog. Your skin is tickled by small impacts, then battered by what feel like hard-thrown marbles. Your arms and legs feel as though they are caught in turbulence, pulling to and fro,

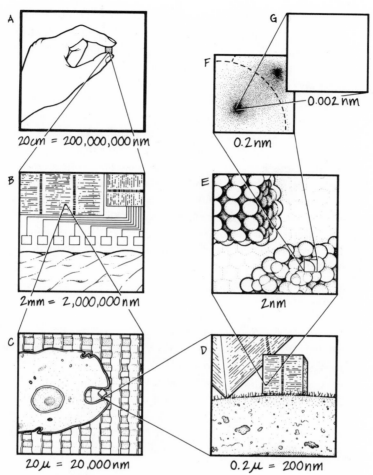

FIGURE 1: POWERS OF TEN

Frame (A) shows a hand holding a computer chip. This is shown magnified 100 times in (B). Another factor of 100 magnification (C) shows a living cell placed on the chip to show scale. Yet another factor of 100 magnification (D) shows two nanocomputers beside the cell. The smaller (shown as a block) has roughly the same power as the chip seen in the first view; the larger (with only the corner visible) is as powerful as a mid-1980s mainframe computer. Another factor of 100 magnification (E) shows an irregular protein from the cell on the lower right, and a cylindrical gear made by molecular manufacturing at top left. Taking a smaller factor of 10 jump, (F) shows two atoms in the protein, with the electron clouds represented by stippling. A final factor of 100 magnification (G) reveals the nucleus of the atom as a tiny speck.

harder and harder. The ground hits your feet, you stumble and stick to the ground like a fly on flypaper, battered so hard that it almost hurts. You asked for realism, and only the built-in safety limits in the suit keep the simulated thermal motions of air molecules and of your own arms from beating you senseless.

"Stop!" gives you a rest from the suit's yanking and thumping, and "Standard settings!" makes the world around you become more reasonable. The simulation changes, introducing the standard cheats. Your simulated eyes are now smaller than a light wave, making focus impossible, but the goggles snap your vision into sharpness and show the atoms around you as small spheres. (Real nanomachines are as blind as you were a moment ago, and can't cheat.) You are on the surface of the 1990s computer chip, between a cell and two blocky nanocomputers like the ones in Figure 1D. Your simulated body is 50 nanometers tall, about 1/40,000,000 your real size, and the smaller nanocomputer is twice your height. At that size, you can "see" atoms and molecules, as in Figure 1E.

The simulation keeps bombarding you with air molecules, but the standard settings leave out the sensation of being pelted with marbles. A moment ago you were stuck tight to the ground by molecular stickiness, but the standard settings give your muscles the effective strength of steel—at least in simulation—by making everything around you much softer and weaker. The tourguide says that the only unreal features of the simulation have to do with you—not just your ability to see and to ignore thermal shaking and bombardment, but also your sheer existence at a size too small for anything so complex as a human being. It also explains why you can see things move, something about slowing down everything around you by a factor of 10 for every factor of 10 enlargement, and by another factor to allow for your being made stronger and hence faster. And so, with your greater strength and some adjustments to make your arms, legs, and torso less sticky, you can stand, see, feel, and take stock of the situation.

MOLECULAR TEXTURE

The ground underfoot, like everything around you, is pebbly with atom-sized bumps the size of your fingertips. Objects look like bunches of transparent grapes or fused marbles in a variety of pretty but imaginary colors. The simulation displays a view of atoms and molecules much like those used by chemists in the 1980s, but with a sharper 3-D image and a better way to move them and to feel the forces they exert. Actually, the whole simulation setup is nothing but an improved version of systems built in the late 1980s—the computer is faster, but it is calculating the same things. The video goggles are better and the whole-body powersuit is a major change, but even in the 1980s there were 3-D displays for molecules and crude devices that gave a sense of touching them.

The gloves on this suit give the sensation of touching whatever the computer simulates. When you run a fingertip over the side of the smaller nanocomputer, it feels odd, hard to describe. It is as if the surface were magnetic—it pulls on your fingertip if you move close enough. But the result isn't a sharp click of contact, because the surface isn't hard like a magnet, but strangely soft. Touching the surface is like touching a film of fog that grades smoothly into foam rubber, then hard rubber, then steel, all within the thickness of a sheet of corrugated cardboard. Moving sideways, your fingertip feels no texture, no friction, just smooth bumps more slippery than oil, and a tendency to get pulled into hollows. Pulling free of the surface takes a firm tug. The simulation makes your atom-sized fingertips feel the same forces that an atom would. It is strange how slippery the surface is—and it can't have been lubricated, since even a single oil molecule would be a lump the size of your thumb. This slipperiness makes it obvious how nano-scale bearings can work, how the parts of molecular machines can slide smoothly.

But on top of this, there is a tingling feeling in your fingers, like the sensation of touching a working loudspeaker. When you put your ear against the wall of the nanocomputer, you flinch back: for a moment, you heard a sound like the hiss of a twentieth-century television tuned to a channel with no broadcast, with nothing but snow and static—but loud, painfully loud. All the atoms in the surface are vibrating at high frequencies, too fast to see. This is thermal vibration, and it's obvious why it's also called thermal noise.

GAS AND LIQUID

Individual molecules still move too quickly to see. So, to add one more cheat to the simulation, you issue the command "Whoa!" and everything around seems to slow down by a factor of ten.

On the surface, you now can see thermal vibrations that had been too quick to follow. All around, air molecules become easier to watch. They whiz about as thick as raindrops in a storm, but they are the size of marbles and bounce in all directions. They're also sticky in a magnetlike way, and some are skidding around on the wall of the nanocomputer. When you grab one, it slips away. Most are like two fused spheres, but you spot one that is perfectly round—it is an argon atom, and these are fairly rare. With a firm grip on all sides to keep it from shooting away like a watermelon seed, you pinch it between your steel-strong fingers. It compresses by about 10 percent before the resistance is more than you can overcome. It springs back perfectly and instantly when you relax, then bounces free of your grip. Atoms have an unfamiliar perfection about them, resilient and unchanging, and they surround you in thick swarms.

At the base of the wall is a churning blob that can only be a droplet of water. Scooping up a handful for a closer look yields a swarm of molecules, hundreds, all tumbling and bumbling over one another, but clinging in a coherent mass. As you watch, though, one breaks free of the liquid and flies off into the freer chaos of the surrounding air: the water is evaporating. Some slide up your arm and lodge in the armpit, but eventually skitter away. Getting rid of all the water molecules takes too much scraping, so you command "Clean me!" to dry off.

TOO SMALL AND TOO LARGE

Beside you, the smaller nanocomputer is a block twice your height, but it's easy to climb up onto it as the tourguide suggests. Gravity is less important on a small scale: even a fly can defy gravity to walk on a ceiling, and an ant can lift what would be a truck to us. At a simulated size of fifty nanometers, gravity counts for nothing. Materials keep their strength, and are just as hard to bend or break, but the weight of an object becomes negligible. Even without the strength-enhancement that lets you overcome molecular stickiness, you could lift an object with 40 million times your mass—like a person of normal size lifting a box containing a half-dozen fully loaded oil tankers. To simulate this weak gravity, the powersuit cradles your body's weight, making you feel as if you were floating. This is almost like a vacation in an orbital theme park, walking with stickyboots on walls, ceilings, and whatnot, but with no need for antinausea medication.

On top of the nanocomputer is a stray protein molecule, like the one in Figure 1E. This looks like a cluster of grapes and is about the same size. It even feels a bit like a bunch of grapes, soft and loose. The parts don't fly free like a gas or tumble and wander like a liquid, but they do quiver like gelatin and sometimes flop or twist. It is solid enough, but the folded structure is not as strong as your steel fingers. In the 1990s, people began to build molecular machinery out of proteins, copying biology. It worked, but it's easy to see why they moved on to better materials.

From a simulated pocket, you pull out a simulated magnifying glass and look at the simulated protein. This shows a pair of bonded atoms on the surface at 10 times magnification, looking like Figure 1F. The atoms are almost transparent, but even a close look doesn't reveal a nucleus inside, because it's too small to see. It would take 1,000 times magnification to be able to see it, even with the head start of being able to see atoms with your naked eye. How could people ever confuse big, plump atoms with tiny specks like nuclei? Remembering how your steel-strong fingers couldn't press more than a fraction of the way toward the nucleus of an argon atom from the air, it's clear why nuclear fusion is so difficult. In fact, the tourguide said that it would take a real-world projectile over a hundred times faster than a high-powered rifle bullet to penetrate into the atomic core and let two nuclei fuse. Try as you might, there just isn't anything you could find in the molecular world that could reach into the middle of an atom to meddle with its nucleus. You can't touch it and you can't see it, so you stop squinting through the magnifying glass. Nuclei just aren't of much interest in nanotechnology.

PUZZLE CHAINS

Taking the advice of the tourguide, you grab two molecular knobs on the protein and pull. It resists for a moment, but then a loop comes free, letting other loops flop around more, and the whole structure seems to melt into a writhing coil. After a bit of pulling and wrestling, the protein's structure becomes obvious: It is a long chain—longer than you are tall, if you could get it straight—and each segment of the chain has one of several kinds of knobs sticking off to the side. With the multicolored, glassy-bead portrayal of atoms, the protein chain resembles a flamboyant necklace. This may be decorative, but how does it all go back together? The chain flops and twists and thrashes, and you pull and push and twist, but the original tight, solid packing is lost. There are more ways to go wrong in folding up the chain than there are in solving Rubik's Cube, and now that the folded structure is gone, it isn't even clear what the result should look like. How did those twentieth-century researchers ever solve the notorious "protein-folding problem"? It's a matter of record that they started building protein objects in the late 1980s.

This protein molecule won't go back together, so you try to break it. A firm grip and a powerful yank straightens a section a bit, but the chain holds together and snaps back. Though unfolding it was easy, even muscles with the strength of steel—the strength of Superman—can't break the chain itself. Chemical bonds are amazingly strong, so it's time to cheat again. When you say, "Flimsy world—one second!" while pulling, your hands easily move apart, splitting the chain in two before its strength returns to normal. You've forced a chemical change, but there must be easier ways since chemists do their work without tiny superhands. While you compare the broken ends, they thrash around and bump together. The third time this happens, the chain rejoins, as strong as before. This is like having snap-together parts, but the snaps are far stronger than welded steel. Modern assembler chemistry usually uses other approaches, but seeing this happen makes the idea of molecular assembly more understandable: Put the right pieces together in the right positions, and they snap together to make a bigger structure.

Remembering the "Whoa!" command, you decide to go back to the properly scaled speed for your size and strength. Saying "Standard settings!," you see the thrashing of the protein chain speed up to a hard-to-follow blur.

NANOMACHINES

At your feet is a ribbed, ringed cylindrical object about the size of a soup can—not a messy, loosely folded strand like the protein (before it fell apart), but a solid piece of modern nanotechnology. It's a gear like the one in Figure 1E. Picking it up, you can immediately feel how different it is from a protein. In the gear, *everything* is held in place by bonds as strong as those that strung together the beads of the protein chain. It can't unfold, and you'd have to cheat again to break its perfect symmetry. Like those in the wall of the nanocomputer, its solidly attached atoms vibrate only slightly. There's another gear nearby, so you fit them together and make the atomic teeth mesh, with bumps on one fitting into hollows on the other. They stick together, and the soft, slick atomic surfaces let them roll smoothly.

Underfoot is the nanocomputer itself, a huge mechanism built in the same rigid style. Climbing down from it, you can see through the transparent layers of the wall to watch the inner works.

An electric motor an arm-span wide spins inside, turning a crank that drives a set of oscillating rods, which in turn drive smaller rods. This doesn't look like a computer; it looks more like an engineer's fantasy from the nineteenth century. But then, it is an antique design—the tourguide said that the original proposal was a piece of exploratory engineering dating from the mid-1980s, a mechanical design that was superseded by improved electronic designs before anyone had the tools to build even a prototype. This simulation is based on a version built by a hobbyist many years later.

The mechanical nanocomputer may be crude, but it does work, and it's a lot smaller and more efficient than the electronic computers of the early 1990s. It's even somewhat faster. The rods slide back and forth in a blur of motion, blocking and unblocking each other in changing patterns, weaving patterns of logic. This nanocomputer is a stripped-down model with almost no memory, useless by itself. Looking beyond it, you see the other block—the one on the left in Figure 1D—which contains a machine powerful enough to compete with most computers built in 1990. This computer is a millionth of a meter on a side, but from where you stand, it looks like a blocky building looming over ten stories tall. The tourguide says that it contains over 100 billion atoms and stores as much data as a room full of books. You can see some of the storage system inside: row upon row of racks containing spools of molecular tape somewhat like the protein chain, but with simple bumps representing the 1s and 0s of computer data.

These nanocomputers seem big and crude, but the ground you're now standing on is also a computer—a single chip from 1990, roughly as powerful as the smaller, stripped-down nanocomputer at your side. As you gaze out over the chip, you get a better sense for just how crude things were a few decades ago. At your feet, on the smallest scale, the chip is an irregular mess. Although the wall of the nanocomputer is pebbly with atomic-scale bumps, the bumps are as regular as tile. The chip's surface, though, is a jumble of lumps and mounds. This pattern spreads for dozens of paces in all directions, ending in an irregular cliff marking the edge of a single transistor. Beyond, you can see other ridges and plateaus stretching off to the horizon. These form grand, regular patterns, the circuits of the computer. The horizon—the edge of the chip— is so distant that walking there from the center would (as the tourguide warns) take *days*. And these vast pieces of landscaping were considered twentieth-century miracles of miniaturization?

CELLS AND BODIES

Even back then, research in molecular biology had revealed the existence of smaller, more perfect machines such as the protein molecules in cells. A simulated human cell—put here because earlier visitors wanted to see the size comparison—sits on the chip next to the smaller nanocomputer. The tourguide points out that the simulation cheats a bit at this point, making the cell act as though it were in a watery environment instead of air. The cell dwarfs the nanocomputer, sprawling across the chip surface and rearing into the sky like a small mountain. Walking the nature trail around its edge would lead across many transistor-plateaus and take about an hour. A glance is enough to show how different it is from a nanocomputer or a gear: it *looks* organic, it bulges and curves like a blob of liver, but its surface is shaggy with waving molecular chains.

Walking up to its edge, you can see that the membrane wrapping the cell is fluid (cell *walls* are for stiff things like plants), and the membrane molecules are in constant motion. On an impulse, you thrust your arm through the membrane and poke around inside. You can feel many proteins bumping and tumbling around in the cell's interior fluid, and a crisscrossing network of protein cables and beams. Somewhere inside are the molecular machines that made all these proteins, but such bits of machinery are embedded in a roiling, organic mass. When you pull your arm out, the membrane flows closed behind. The fluid, dynamic structure of the cell is largely self-healing. That's what let scientists perform experimental surgery on cells with the old, crude tools of the twentieth century: They didn't need to stitch up the holes they made when they poked around inside.

Even a single human cell is huge and complex. No real thinking being could be as small as you are in the simulation: A simple computer without any memory is twice your height, and the larger nanocomputer, the size of an apartment complex, is no smarter than one of the submoronic computers of 1990. Not even a bendable finger could be as small as your simulated fingers: in the simulation, your fingers are only one atom wide, leaving no room for the slimmest possible tendon, to say nothing of nerves.

For a last look at the organic world, you gaze out past the horizon and see the image of your own, full-sized thumb holding the chip on which you stand. The bulge of your thumb rises ten times higher than Mount Everest. Above, filling the sky, is a face looming like the Earth seen from orbit, gazing down. It is your face, with cheeks the size of continents. The eyes are motionless. Thinking of the tourguide's data, you remember: The simulation uses the standard mechanical scaling rules, so being 40 million times smaller has made you 40 million times faster. To let you pull free of surfaces, it increases your strength by more than a factor of 100, which increased your speed by more than a factor of 10. So one second in the ordinary world corresponds to over 400 million here in the simulation. It would take years to see that huge face in the sky complete a single eyeblink.

Enough. At the command "Get me out!", the molecular world vanishes, and your feeling of weight returns as the suit goes slack. You strip off the video goggles—and hugely, slowly, blink.

THREE

Bottom-Up Technology

T he tour in the last chapter showed the sizes, forces, and general nature of objects in the molecular world. Building on this, we can get a better picture of where developments seem to be leading, a better picture of molecular manufacturing itself. To show the sizes, forces, and general nature of things in molecular manufacturing, we first invite the reader (and the reader's inquisitive alter ego) to take a second and final tour before returning to the world of present-day research. As before, the pre-1990 history is accurate, and the science isn't fiction.

THE SILICON VALLEY FAIRE

The tour of the molecular world showed some products of molecular manufacturing, but didn't show how they were made. The technologies you remember from the old days have mostly been replaced—but how did this happen? The Silicon Valley Faire is advertised as "An authentic theme park capturing life, work, and play in the early Breakthrough years." Since "work" must include manufacturing, it seems worth a visit.

A broad dome caps the park—"To fully capture the authentic sights, sounds, and smells of the era," the tourguide politely says. Inside, the clothes and hairstyles, the newspaper headlines, the bumper-to-bumper traffic, all look much as they did before your long nap. A light haze obscures the buildings on the far side of the dome, your eyes burn slightly, and the air smells truly authentic.

POCKET LIBRARIES

The Nanofabricators, Inc., plant offers the main display of early nanotechnology. As you near the building, the tourguide mentions that this is indeed the original manufacturing plant, given landmark status over twenty years ago, then made the centerpiece of the Silicon Valley Faire ten years later, when . . . With a few taps, you reset the pocket tourguide to speak up less often.

As people file into the Nanofabricator plant, there's a moment of hushed quiet, a sense of walking into history. Nanofabricators: home of the SuperChip, the first mass-market product of nanotechnology. It was the huge memory capacity of SuperChips that made possible the first Pocket Library.

This section of the plant now houses a series of displays, including working replicas of early products. Picking up a Pocket Library, you find that it's not only the size of a wallet, but about the same weight. Yet it has enough memory to record every volume in the Library of Congress—something like a million times the capacity of a personal computer from 1990. It opens with a flip, the two-panel screen lights up, and a world of written knowledge is at your fingertips. Impressive.

"Wow, can you believe these things?" says another tourist as he fingers a Pocket Library. "Hardly any video, no 3-D—just words, sound, and flat pictures. And the cost! I wouldn't've bought 'em for my kids at *that* price!"

Your tourguide quietly states the price: about what you remember for a top-of-the-line TV set from 1990. This isn't the cheap manufacturing promised by mature nanotechnology, but it seems like a pretty good price for a library. Hmm . . . how did they work out the copyrights and royalties? There's a lot more to this product than just the technology. . . .

NANOFABRICATION

The next room displays more technology. Here in the workroom where SuperChips were first made, early nanotech manufacturing is spread out on display. The whole setup is surprisingly quiet and ordinary. Back in the 1980s and 1990s, chip plants had carefully controlled clean rooms with gowns and masks on workers and visitors, special workstations, and carefully crafted air flows to keep dust away from products. This room has none of that. It's even a little grubby.

In the middle of a big square table are a half-dozen steel tanks, about the size and shape of old-fashioned milk cans. Each can has a different label identifying its contents: MEMORY BLOCKS, DATA-TRANSMISSION BLOCKS, INTERFACE BLOCKS. These are the parts needed for building up the chip. Clear plastic tubes, carrying clear and tea-colored liquids, emerge from the mouths of the milk cans and drape across the table. The tubes end in fist-sized boxes mounted above shallow dishes sitting in a ring around the cans. As the different liquids drip into each dish, a beater like a kitchen mixer swirls the liquid. In each dish, nanomachines are building SuperChips.

A Nanofab "engineer," dressed in period clothing complete with name badge, is setting up a dish to begin building a new chip. "This," he says, holding up a blank with a pair of tweezers, "is a silicon chip like the ones made with pre-breakthrough technology. Companies here in this valley made chips like these by melting silicon, freezing it into lumps, sawing the lumps into slices, polishing the slices, and then going through a long series of chemical and photographic steps. When they were done, they had a pattern of lines and blobs of different materials on the surface. Even the smallest of these blobs contained *billions* of atoms, and it took several blobs working together to store a single bit of information. A

chip this size, the size of your fingernail, could store only a fraction
of a billion bits. Here at Nanofab, we used bare silicon chips as a
base for building up nanomemory. The picture on the wall here
shows the surface of a blank chip: no transistors, no memory circuits, just fine wires to connect up with the nanomemory we built
on top. The nanomemory, even in the early days, stored *thousands*
of billions of bits. And we made them like this, but a thousand at a
time—" He places the chip in the dish, presses a button, and the
dish begins to fill with liquid.

"A few years later," he adds, "we got rid of the silicon chips
entirely"—he props up a sign saying THIS CHIP BUILD BEGAN AT: 2:15
P.M., ESTIMATED COMPLETION TIME: 1:00 A.M.—"and we sped up the
construction process by a factor of a thousand."

The chips in the dishes all look pretty much the same except
for color. The new chip looks like dull metal. The only difference
you can see in the older chips, further along in the process, is a
smooth rectangular patch covered by a film of darker material. An
animated flowchart on the wall shows how layer upon layer of nanomemory building blocks are grabbed from solution and laid down
on the surface to make that film. The tourguide explains that the
energy for this process, like the energy for molecular machines
within cells, comes from dissolved chemicals—from oxygen and
fuel molecules. The total amount of energy needed here is trivial,
because the amount of product is trivial: at the end of the process,
the total thickness of nanomemory structure—the memory store for
a Pocket Library—amounts to one-tenth the thickness of a sheet of
paper, spread over an area smaller than a postage stamp.

MOLECULAR ASSEMBLY

The animated flowchart showed nanomemory building blocks as
big things containing about a hundred thousand atoms apiece (it
takes a moment to remember that this is still submicroscopic). The
build process in the dishes stacked these blocks to make the memory film on the SuperChip, but how were the blocks themselves
built? The hard part of this molecular-manufacturing business has
got to be at the bottom of the whole process, at the stage where

molecules are put together to make large, complex parts.

The Silicon Valley Faire offers simulations of this molecular assembly process, and at no extra charge. From the tourguide, you learn that modern assembly processes are complex; that earlier processes—like those used by Nanofabricators, Inc.—used clever-but-obscure engineering tricks; and that the simplest, earliest concepts were never built. Why not begin at the beginning? A short walk takes you to the Museum of Antique Concepts, the first wing of the Museum of Molecular Manufacturing.

A peek inside the first hall shows several people strolling around wearing loosely fitting jumpsuits with attached goggles and gloves, staring at nothing and playing mime with invisible objects. Oh well, why not join the fools' parade? Stepping through the doorway while wearing the suit is entirely different. The goggles show a normal world outside the door and a molecular world inside. Now you, too, can see and feel the exhibit that fills the hall. It's much like the earlier simulated molecular world: it shares the standard settings for size, strength, and speed. Again, atoms seem 40 million times larger, about the size of your fingertips. This simulation is a bit less thorough than the last was—you can feel simulated objects, but only with your gloved hands. Again, everything seems to be made of quivering masses of fused marbles, each an atom.

"Welcome," says the tourguide, "to a 1990 concept for a molecular-manufacturing plant. These exploratory engineering designs were never intended for actual use, yet they demonstrate the basics of molecular manufacturing: making parts, testing them, and assembling them."

Machinery fills the hall. Overall, the sight is reminiscent of an automated factory of the 1980s or 1990s. It seems clear enough what must be going on: Big machines stand beside a conveyor belt loaded with half-finished-looking blocks of some material (this setup looks much like Figure 2); the machines must do some sort of work on the blocks. Judging by the conveyor belt, the blocks eventually move from one arm to the next until they turn a corner and enter the next hall.

Since nothing is real, the exhibit can't be damaged, so you walk up to a machine and give it a poke. It seems as solid as the wall of the nanocomputer in the previous tour. Suddenly, you notice something odd: no bombarding air molecules and no droplets of

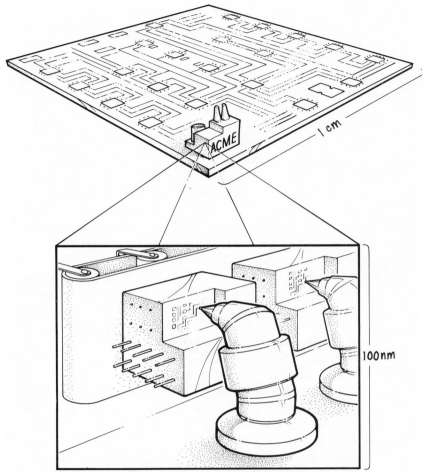

FIGURE 2: ASSEMBLER WITH FACTORY ON CHIP

A factory large enough to make over 10 million nanocomputers per day would fit on the edge of one of today's integrated circuits. Inset shows an assembler arm together with a workpiece on a conveyor belt.

water—in fact, no loose molecules anywhere. Every atom seems to be part of a mechanical system, quivering with thermal vibration, but otherwise perfectly controlled. Everything here is like the nanocomputer or like the tough little gear; none of it resembles the loosely coiled protein or the roiling mass of the living cell.

The conveyor belt seems motionless. At regular intervals along the belt are blocks of material under construction: workpieces. The nearest block is about a hundred marble-bumps wide, so it must contain something like $100 \times 100 \times 100$ atoms, a full million. This block looks strangely familiar, with its rods, crank, and the rest. It's a nanocomputer—or rather, a blocklike part of a nanocomputer still under construction.

Standing alongside the pieces of nanocomputer on the conveyor belt, dominating the hall, is a row of huge mechanisms. Their trunks rise from the floor, as thick as old oaks. Even though they bend over, they rear overhead. "Each machine," your tourguide says, "is the arm of a general-purpose molecular assembler."

One assembler arm is bent over with its tip pressed to a block on the conveyor belt. Walking closer, you see molecular assembly in action. The arm ends in a fist-sized knob with a few protruding marbles, like knuckles. Right now, two quivering marbles— atoms—are pressed into a small hollow in the block. As you watch, the two spheres shift, snapping into place in the block with a quick twitch of motion: a chemical reaction. The assembler arm just stands there, nearly motionless. The fist has lost two knuckles, and the block of nanocomputer is two atoms larger.

The tourguide holds forth: "This general-purpose assembler concept resembles, in essence, the factory robots of the 1980s. It is a computer-controlled mechanical arm that moves molecular tools according to a series of instructions. Each tool is like a single-shot stapler or rivet gun. It has a handle for the assembler to grab and comes loaded with a little bit of matter—a few atoms—which it attaches to the workpiece by a chemical reaction." This is like the rejoining of the protein chain in the earlier tour.

MOLECULAR PRECISION

The atoms seemed to jump into place easily enough; can they jump out of place just as easily? By now the assembler arm has crept back from the surface, leaving a small gap, so you can reach in and poke at the newly added atoms. Poking and prying do no good: When you push as hard as you can (with your simulated fingers as strong as steel), the atoms don't budge by a visible amount. Strong molecular bonds hold them in place.

Your pocket tourguide—which has been applying the power of a thousand 1990s supercomputers to the task of deciding when to speak up—remarks, "Molecular bonds hold things together. In strong, stable materials, atoms are either bonded, or they aren't, with no possibilities in between. Assemblers work by making and breaking bonds, so each step either succeeds perfectly or fails completely. In pre-breakthrough manufacturing, parts were always made and put together with small inaccuracies. These could add up to wreck product quality. At the molecular scale, these problems vanish. Since each step is perfectly precise, little errors can't add up. The process either works, or it doesn't."

But what about those definite, complete failures? Fired by scientific curiosity, you walk to the next assembler, grab the tip, and shake it. Almost nothing happens. When you shove as hard as you can, the tip moves by about one tenth of an atomic diameter, then springs back. "Thermal vibrations can cause mistakes by causing parts to come together and form bonds in the wrong place," the tourguide remarks. "Thermal vibrations make floppy objects bend further than stiff ones, and so these assembler arms were designed to be thick and stubby to make them very stiff. Error rates can be kept to one in a trillion, and so small products can be perfectly regular and perfectly identical. Large products can be *almost* perfect, having just a few atoms out of place." This should mean high reliability. Oddly, most of the things you've been seeing outside have looked pretty ordinary—not slick, shiny, and perfect, but rough and homey. They must have been *manufactured* that way, or made by hand. Slick, shiny things must not impress anyone anymore.

MOLECULAR ROBOTICS

By now, the assembler arm has moved by several atom-widths. Through the translucent sides of the arm you can see that the arm is full of mechanisms: twirling shafts, gears, and large, slowly turning rings that drive the rotation and extension of joints along the trunk. The whole system is a huge, articulated robot arm. The arm is big because the smallest parts are the size of marbles, and the machinery inside that makes it move and bend has many, many parts. Inside, another mechanism is at work: The arm now ends in a hole, and you can see the old, spent molecular tool being retracted through a tube down the middle.

Patience, patience. Within a few minutes, a new tool is on its way back up the tube. Eventually, it reaches the end. Shafts twirl, gears turn, and clamps lock the tool in position. Other shafts twirl, and the arm slowly leans up against the workpiece again at a new site. Finally, with a twitch of motion, more atoms jump across, and the block is again just a little bit bigger. The cycle begins again. This huge arm seems amazingly slow, but the standard simulation settings have shifted speeds by a factor of over 400 million. A few minutes of simulation time correspond to less than a millionth of a second of real time, so this stiff, sluggish arm is completing about a million operations per second.

Peering down at the very base of the assembler arm, you can get a glimpse of yet more assembler-arm machinery underneath the floor: Electric motors spin, and a nanocomputer chugs away, rods pumping furiously. All these rods and gears move quickly, sliding and turning many times for every cycle of the ponderous arm. This seems inefficient; the mechanical vibrations must generate a lot of heat, so the electric motors must draw a lot of power. Having a computer control each arm is a lot more awkward now than it was in pre-breakthrough years. Back then, a robot arm was big and expensive and a computer was a cheap chip; now the computer is bigger than the arm. There must be a better way—but then, this *is* the Museum of Antique Concepts.

BUILDING BLOCKS INTO BUILDINGS

Where do the blocks go, once the assemblers have finished with them? Following the conveyor belt past a dozen arms, you stroll to the end of the hall, turn the corner, and find yourself on a balcony overlooking a vaster hall beyond. Here, just off the conveyor belt, a block sits in a complex fixture. Its parts are moving, and an enormous arm looms over it like a construction crane. After a moment,

the tourguide speaks up and confirms your suspicion: "After manufacturing, each block is tested. Large arms pick up properly made
blocks. In this hall, the larger arms assemble almost a thousand
blocks of various kinds to make a complete nanocomputer."

The grand hall has its own conveyor belt, bearing a series of
partially completed nanocomputers. Arrayed along this grand belt is
a row of grand arms, able to swing to and fro, to reach down to
lesser conveyor belts, pluck million-atom blocks from testing sta-
tions, and plug them into the grand workpieces, the nanocomputers
under construction. The belt runs the length of the hall, and at the
end, finished nanocomputers turn a corner—to a yet-grander hall
beyond?

After gazing at the final-assembly hall for several minutes, you
notice that nothing seems to have moved. Mere patience won't do:
at the rate the smaller arms moved in the hall behind you, each
block must take months to complete, and the grand block-handling
arms are taking full advantage of the leisure this provides. Building
a computer, start to finish, might take a terribly long time. Perhaps
as long as the blink of an eye.

Molecular assemblers build blocks that go to block assemblers.
The block assemblers build computers, which go to system assemblers, which build systems, which—at least one path from molecules to large products seems clear enough. If a car were assembled
by normal-sized robots from a thousand pieces, each piece having
been assembled by smaller robots from a thousand smaller pieces,
and so on, down and down, then only ten levels of assembly process would separate cars from molecules. Perhaps, around a few
more corners and down a few more ever-larger halls, you would see
a post-breakthrough car in the making, with unrecognizable engine
parts and comfortable seating being snapped together in a century-
long process in a hall so vast that the Pacific Ocean would be a
puddle in the corner. . . .

Just ten steps in size; eight, starting with blocks as big as the
ones made in the hall behind you. The molecular world seems
closer, viewed this way.

MOLECULAR PROCESSING

Stepping back into that hall, you wonder how the process begins. In every cycle of their sluggish motion, each molecular assembler gets a fresh tool through a tube from somewhere beneath the floor, and that *somewhere* is where the story of molecular precision begins. And so you ask, "Where do the tools come from?" and the tourguide replies, "You might want to take the elevator to your left."

Stepping out of the elevator and into the basement, you see a wide hall full of small conveyor belts and pulleys; a large pipe runs down the middle. A plaque on the wall says, "Mechanochemical processing concept, circa 1990." As usual, all the motions seem rather slow, but in this hall everything that seems designed to move is visibly in motion. The general flow seems to be away from the pipe, through several steps, and then up through the ceiling toward the hall of assemblers above.

After walking over to the pipe, you can see that it is nearly transparent. Inside is a seething chaos of small molecules: the wall of the pipe is the boundary between loose molecules and controlled ones, but the loose molecules are well confined. In this simulation, your fingertips are like small molecules. No matter how hard you push, there's no way to drive your finger through the wall of the pipe. Every few paces along the pipe a fitting juts out, a housing with a mechanically driven rotating thing, exposed to the liquid inside the pipe, but also exposed to a belt running over one of the pulleys, embedded in the housing. It's hard to see exactly what is happening.

The tourguide speaks up, saying, "Pockets on the rotor capture single molecules from the liquid in the pipe. Each rotor pocket has a size and shape that fits just one of the several different kinds of molecule in the liquid, so the process is rather selective. Captured molecules are then pushed into the pockets on the belt that's wrapped over the pulley there, then—"

"Enough," you say. Fine, it singles out molecules and sticks them into this maze of machinery. Presumably, the machines can sort the molecules to make sure the right kinds go to the right places.

The belts loop back and forth carrying big, knobby masses of molecules. Many of the pulleys—rollers?—press two belts together inside a housing with auxiliary rollers. While you are looking at one of these, the tourguide says, "Each knob on a belt is a mechanochemical-processing device. When two knobs on different belts are pressed together in the right way, they are designed to transfer molecular fragments from one to another by means of a mechanically forced chemical reaction. In this way, small molecules are broken down, recombined, and finally joined to molecular tools of the sort used in the assemblers in the hall above. In this device here, the rollers create a pressure equal to the pressure found halfway to the center of the Earth, speeding a reaction that—"

"Fine, fine," you say. Chemists in the old days managed to make amazingly complex molecules just by mixing different chemicals together in solution in the right order under the right conditions. Here, molecules can certainly be brought together in the right order, and the conditions are much better controlled. It stands to reason that this carefully designed maze of pulleys and belts can do a better job of molecule processing than a test tube full of disorganized liquid ever could. From a liquid, through a sorter, into a mill, and out as tools: this seems to be the story of molecule processing. All the belts are loops, so the machinery just goes around and around, carrying and transforming molecular parts.

BEYOND ANTIQUES

This system of belts seems terribly simple and efficient, compared to the ponderous arms driven by frantic computers in the hall above. Why stop with making simple tools? You must have muttered this, because the tourguide speaks up again and says, "The Special-Assembler Exhibit shows another early molecular-manufacturing concept that uses the principles of this molecule-processing system to build large, complex objects. If a system is building only a single product, there is no need to have computers and flexible arms move parts around. It is far more efficient to build a machine in which everything just moves on belts at a constant speed, adding small parts to larger ones and then bringing the larger ones together as you saw at the end of the hall above."

This does seem like a more sensible way to churn out a lot of identical products, but it sounds like just more of the same. Gears like fused marbles, belts like coarse beadwork, drive shafts, pulleys, machines and more machines. In a few places, marbles snap into new patterns to prepare a tool or make a product. Roll, roll, chug, chug, pop, snap, then roll and chug some more.

As you leave the simulation hall, you ask, "Is there anything important I've missed in this molecular manufacturing tour?"

The tourguide launches into a list: "Yes—the inner workings of assembler arms, with drive shafts, worm gears, and harmonic drives; the use of Diels-Adler reactions, interfacial free-radical chain reactions, and dative-bond formation to join blocks together in the larger-scale stages of assembly; different kinds of mechanochemical processing for preparing reactive molecular tools; the use of staged-cascade methods in providing feed-molecules of the right kinds with near-perfect reliability; the differences between efficient and inefficient steps in molecular processing; the use of redundancy to ensure reliability in large systems despite sporadic damage; modern methods of building large objects from smaller blocks; modern electronic nanocomputers; modern methods for—"

"Enough!" you say, and the tourguide falls silent as you pitch it into a recycling bin. A course in molecular manufacturing isn't what you're looking for right now; the general idea seems clear enough. It's time to take another look at the world on a more normal scale. Houses, roads, buildings, even the landscape, looked different out there beyond the Faire dome—less crowded, paved, and plowed than you remember. But why? The history books (well, they're more than just *books*) say that molecular manufacturing made a big difference; perhaps now the changes will make more sense. Yes, it's time to leave.

As you toss your goggled, gloved jumpsuit into another bin, a striking dark-haired woman is taking a fresh one from a rack. She wears a jacket emblazoned with the name "Desert Rose Nano-Manufacturing."

"How'd you like it?" she asks with a smile.

"Pretty amazing," you say.

"Yes," she agrees. "I saw this sim back when I was taking my first molecular-manufacturing class. I swore I'd never design anything so clunky! This whole setup really brings back the memories—I can't wait to see if it's as crude as I remember." She steps into the simulation hall and closes the door.

CRUDE TECHNOLOGY

As the Silicon Valley Faire scenario shows, molecular manufacturing will work much like ordinary manufacturing, but with devices built so small that a single loose molecule of pollutant would be like a brick heaved into a machine tool. John Walker of Autodesk, a leading company in computer-aided design, observes that nanotechnology and today's crude methods are very different: "Technology has never had this kind of precise control; all of our technologies today are bulk technologies. We take a big chunk of stuff and hack away at it until we're left with the object we want, or we assemble parts from components without regard to structure at the molecular level."

Molecular manufacturing will orchestrate atoms into products of symphonic complexity, but modern manufacturing mostly makes loud noises. These figurative noises are sometimes all too literal: A crack in a metal forging grows under stress, a wing fails, and a passenger jet crashes from the sky. A chemical reaction goes out of control, heat and pressure build, and a poisonous blast shakes the countryside. A lifesaving product cannot be made, a heart fails, and a hospital's heart-monitoring machine signals the end with a high-pitched wail.

Today, we make many things from metal, by machining. From the perspective of our standard, simulated molecular world, a typical metal part is a piece of terrain many days' journey across. The metal itself is weak compared to the bonds of the protein chain or other tough nanomechanisms: solid steel is no stronger than your simulated fingers, and the atoms on its surface can be pushed around with your bare hands. Standing on a piece of metal being machined in a lathe, you would see a cutting blade crawl past a few times per years, like a majestic plough the size of a mountain range. Each pass would rip up a strip of the metal landscape, leaving a rugged valley broad enough to hold a town. This is machining from a nanotechnological perspective: a process that hacks crude shapes from intrinsically weak materials.

Today, electronics are made from silicon chips. We have already seen the landscape of a finished chip. During manufacturing, metal features would be built up by a centuries-long drizzle of metal-atom rain, and hollows would be formed by a centuries-long submergence in an acid sea. From the perspective of our simulation, the whole process would resemble geology as much as manufacturing, with the slow layering of sedimentary deposits alternating with ages of erosion. The term *nanotechnology* is sometimes used as a name for small-scale microtechnology, but the difference between molecular manufacturing and this sort of microlandscaping is like the difference between watchmaking and bulldozing.

Today, chemists make molecules by solution chemistry. We have seen what a liquid looks like in our first simulation, with molecules bumping and tumbling and wandering around. Just as assemblers can make chemical reactions occur by bringing molecules together mechanically, so reactions can occur when molecules bump at random through thermal vibration and motion in a liquid. Indeed, much of what we know today about chemical reactions comes from observing this process. Chemists make large molecules by mixing small molecules in a liquid. By choosing the right molecules and conditions, they can get a surprising measure of control over the results: only some pairs of molecules will react, and then only in certain ways.

Doing chemistry this way, though, is like trying to assemble a model car by putting the pieces in a box and shaking. This will only work with cleverly shaped pieces, and it is hard to make anything very complex. Chemists today consider it challenging to make a precise, three-dimensional structure having a hundred atoms, and making one with a thousand atoms is a great accomplishment. Molecular manufacturing, in contrast, will routinely assemble millions or billions. The basic chemical principles will be the same, but control and reliability will be vastly greater. It is the difference between throwing things together blindly and putting them together with a watchmaker's care.

Technology today doesn't permit thorough control of the structure of matter. Molecular manufacturing will. Today's technologies have given us computers, spacecraft, indoor plumbing, and the other wonders of the modern age. Tomorrow's will do much more, bringing change and choice.

SIMPLE MATTER, SMART MATTER

Today's technology mostly works with matter in a few basic forms: gases, liquids, and solids. Though each form has many varieties, all are comparatively simple.

Gases, as we've seen, consist of molecules ricocheting through space. A volume of gas will push against its walls and, if not walled in, expand without limit. Gases can supply certain raw materials for nanomachines, and nanomachines can be used to remove pollutants from air and turn them into something else. Gases lack structure, so they will remain simple.

Liquids are somewhat like gases, but their molecules cling together to form a coherent blob that won't expand beyond a certain limit. Liquids will be good sources of raw materials for nanomachines because they are denser and can carry a wide range of fuels and raw materials in solution (the pipe in the molecular-processing hall contained liquid). Nanomachines can clean up polluted water as easily as air, removing and transforming noxious molecules. Liquids have more structure than gases, but nanotechnology will have its greatest application to solids.

Solids are diverse. Solid butter consists of molecules stronger than steel, but the molecules cling to one another by the weaker forces of molecular stickiness. A little heat increases thermal vibrations and makes the solid structure disintegrate into a blob of liquid. Butterlike materials would make poor nanomachines. Metals consist of atoms

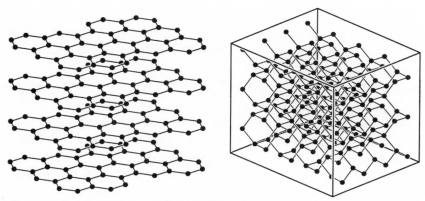

FIGURE 3: CARBON—SOFT AND HARD K. ERIC DREXLER

On the left is graphite—the material called "lead" in pencils—made of carbon atoms. On the right is diamond—the same atoms arranged in a different pattern.

held together by stronger forces, and so they can be structurally stronger and able to withstand higher temperatures. The forces are not very directional, though, and so planes of metal atoms can slip past one another under pressure; this is why spoons bend, rather than break. This ability to slip makes metals less brittle and easier to shape (with crude technology), but it also weakens them. Only the strongest, hardest, highest-melting-point metals are worth considering as parts of nanomachines.

Diamond consists of carbon atoms held together by strong, directional bonds, like the bonds down the axis of a protein chain (see Figure 3). These directional bonds make it hard for planes of atoms to slip past one another, making diamond (and similar materials) very strong indeed—ten to a hundred times stronger than steel. But the planes can't easily slip, so when the material fails, it doesn't bend, it breaks. Tiny cracks can easily grow, making a large object seem weak. Glass is a similar material: glass windows seem weak—and a scratch makes glass far weaker—yet thin, perfect glass fibers are widely used to make composite materials stronger and lighter than steel. Nanotechnology will be able to build with diamond and similar strong materials, making small, flawless fibers and components.

In engineering today, diamond is just beginning to be used. Japan has pioneered a technology for making diamond at low pressure, and a Japanese company sells a speaker with excellent high-frequency response—the speaker cone is reinforced with a light, stiff film of diamond. Diamond is extraordinary stuff, made from cheap materials like natural gas. U.S. companies are scrambling to catch up.

All these materials are simple. More complex structures lead to more complex properties, and begin to give some hint of what molecular manufacturing will mean for materials.

What if you strung carbon atoms in long chains with side-groups, a bit like a protein chain, and linked them into a big three-dimensional mesh? If the chains were kinked so that they couldn't pack tightly, they would coil up and flop around almost like molecules in a liquid, yet the strong bonds would keep the overall mesh intact. Pulling the whole network would tend to straighten the chains, but their writhing motions would tend to coil them back up. This sort of network has been made: it is called rubber.

Rubber is weak mostly because the network is irregular. When stretched, first one chain breaks, then another, because they don't all become taut at the same time to share and divide the load. A more regular mesh would be as soft as rubber at first, but when stretched to the limit would become stronger than steel. Molecular manufacturing could make such stuff.

The natural world contains a host of good materials—cellulose and lignin in wood, stronger-than-steel proteins in spider's silk, hard ceramics in grains of sand, and more. Many products of molecular manufacturing will be designed for great durability, like sand. Others will be designed to fall apart easily for easy recycling, like wood. Some may be designed for uses where they may be thrown away. In this last category, nanotailored biodegradables will shine. With care, almost any sort of product from a shoe to computer-driven nanomachines can be made to last for a good long time, and then unzip fairly rapidly and very thoroughly into molecules and other bits of stuff all of kinds normally found in the soil.

This gives only a hint of what molecular manufacturing will make possible by giving better control of the structure of solid matter. The most impressive applications will not be superstrong structural materials, improved rubber, and simple biodegradable materials: these are uniform, repetitive structures not greatly different from ordinary materials. These materials are "stupid." When pushed, they resist, or they stretch and bounce back. If you shine light on them, they transmit it, reflect it, or absorb it. But molecular manufacturing can do much more. Rather than heaping up simple molecules, it can build material from trillions of motors, ratchets, light-emitters, and computers.

Muscle is smarter than rubber because it contains molecular machines: it can be told to contract. The products of molecular manufacturing can include materials able to change shape, color, and other properties on command. When a dust mote can contain a supercomputer, materials can be made smart, medicine can be made sophisticated, and the world will be a different place. Smart materials will be discussed in Chapter 8.

IDEAS AND CRITICISMS

We've just seen a picture of molecular manufacturing (of one sort) and of what it can do (in sketchy outline). Now let's look at the idea of nanotechnology itself: Where did it come from, and what do the experts think of it? The next chapter will have more to say on the latter point, presenting the thoughts of researchers who are advancing the field through their own work.

ORIGINS

The idea of molecular nanotechnology, like most ideas, has roots stretching far back in time. In ancient Greece, Democritus suggested that the world was built of durable, invisible particles—atoms, the building blocks of solid objects, liquids, and gases. In the last hundred years, scientists have learned more and more about these building blocks, and chemists have learned more and more ways to combine them to make new things. Decades ago, biologists found molecules that do complex things; they termed them "molecular machines."

Physicist Richard Feynman was a visionary of miniaturization who pointed toward something like molecular nanotechnology: on December 29, 1959, in an after-dinner talk at the annual meeting of the American Physical Society, he proposed that large machines could be used to make smaller machines, which could make still smaller ones, working in a top-down fashion from the macroscale to the microscale. At the end of his talk, he painted a vision of moving individual atoms, pointing out, "The principles of physics, as far as I can see, do not speak against the possibility of maneuvering things atom by atom." He pictured making molecules, pointing clearly in the direction taken by the modern concept of nanotechnology: "But it is interesting that it would be, in principle, possible (I think) for a physicist to synthesize any chemical substance the chemist writes down. Give the orders, and the physicist synthesizes it. How? Put the atoms down where the chemist says, and so you make the substance."

Despite this clear signpost pointing to a potentially revolutionary area, no one filled the conceptual gap between miniature machines and chemical substances. There was no clear concept of making molecular machines able to build more such machines, no notion of controllable molecular manufacturing. With hindsight, one wonders why the gap took so long to fill. Feynman himself didn't follow it up, saying that the ability to maneuver atoms one by one "will really be useless" since chemists would come up with traditional, bulk-process ways to make new chemical substances. For a researcher whose main interest was physics, he had contributed much just by placing the signpost: it was up to others to move forward. Instead, the idea of molecular machines for molecular manufacturing didn't appear for decades.

From today's viewpoint, molecular nanotechnology looks more like an extension of chemistry than like an extension of miniaturization. A mechanical engineer, looking at nanotechnology, might ask, "How can machines be made so small?" A chemist, though, would ask, "How can molecules be made so large?" The chemist has the better question. Nanotechnology isn't primarily about miniaturizing machines, but about extending precise control of molecular structure to larger and larger scales. Nanotechnology is about making (precise) things *big*.

MACROSCOPIC AND MOLECULAR COMPONENTS		
Technology	Function	Molecular Examples
struts, beams, casings	transmit force, hold positions	cell walls, microtubules
cables	transmit tension	collagen, silk
fasteners, glue	connect parts	intermolecular forces
solenoids, actuators	move things	muscle actin, myosin
motors	turn shafts	flagellar motor
drive shafts	transmit torque	bacterial flagella
bearings	support moving parts	single bonds
clamps	hold workpieces	enzymatic binding sites
tools	modify workpieces	enzymes, reactive molecules
production lines	construct devices	enzyme systems, ribosomes
numerical control systems	store and read programs	genetic system

Adapted from K. E. Drexler, *Proceedings of the National Academy of Sciences*, Vol. 78 (1981) pp. 5275–78.

Nature gives the most obvious clues to how this can be done, and it was the growing scientific literature on natural molecular machines that led one of the present authors (Drexler) to propose molecular nanotechnology of the sort described here. A strategy to reach the goal was part of the concept: Build increasingly complex molecular machinery from simpler pieces, including molecular machines able to build more molecular machines. And the motivation for studying this, and publishing? Largely the fear of living in a world that might rush into the new technology blindly, with ugly consequences.

This concept and initial exploratory work started in early 1977 at MIT; the first technical publication came in 1981 in the *Proceedings of the National Academy of Sciences*. For years, MIT remained the center of thinking on nanotechnology and molecular manufacturing: in 1985, the MIT Nanotechnology Study Group was formed; it soon initiated an annual lecture series which grew into a two-day symposium by 1990.

The first book on the topic, *Engines of Creation*, was published in 1986. In 1988, Stanford University became the first to offer a course in molecular nanotechnology, sponsored by the Department of Computer Science. In 1989, this department hosted the first major conference on the subject, cosponsored by the Foresight Institute and Global Business Network. With the upcoming publication of a technical book describing nanotechnology—from molecular mechanical and quantum-mechanical principles up to assembly systems and products—the subject will be easier to teach, and more college courses will become available.

In parallel with the development and spread of ideas about nanotechnology and molecular manufacturing—ideas that remain pure theory, however well grounded—scientists and engineers, working in laboratories to build real tools and capabilities, have been pioneering roads to nanotechnology. Research has come a long way since the mid-1980s, as we'll see in the next chapter. But, as one might expect with a complex new idea that, if true, disrupts a lot of existing plans and expectations, some objections have been heard.

"IT WON'T WORK."

Life might be much simpler if these ideas about nanotechnology had some fatal flaw. If only molecules couldn't be used to form machines, or the machines couldn't be used to build things, then we might be able to keep right on going with our crude technologies: our medicine that doesn't heal, our spacecraft that don't open a new frontier, our oil crises, our pollution, and all the limits that keep us from trading familiar problems for strange ones. Most new ideas are wrong, especially if they purport to bring radical changes. It is not unreasonable to hope that these are wrong. From years of discussions with chemists, physicists, and engineers, it is possible to compile what seems to be a complete list of *basic*, critical questions about whether nanotechnology will work. The questioners generally seem satisfied with the answers.

"WILL THERMAL VIBRATIONS MESS THINGS UP?"

The earlier scenarios describe the nature of thermal vibration and the problems it can cause. Designing nanomachines strong enough and stiff enough to operate reliably despite thermal vibration is a genuine engineering challenge. But calculating the design requirements usually requires only simple textbook principles, and these requirements can be met for everything described in this book.

"WILL QUANTUM UNCERTAINTY MESS THINGS UP?"

Quantum mechanics says that particles must be described as small smears of probability, not as points with perfectly defined locations. This is, in fact, why the atoms and molecules in the simulations felt so soft and smooth: their electrons are smeared out over the whole volume of the molecule, and these electron clouds taper off smoothly and softly toward the edges. Atoms themselves are a bit uncertain in position, but this is a small effect compared to thermal vibrations. Again, simple textbook principles apply, and well-designed molecular machines will work.

"WILL LOOSE MOLECULES MESS THINGS UP?"

Chemists work with loose molecules in liquids, and they naturally tend to picture molecules as flying around loose. It is possible to build nanomachines and molecular manufacturing systems that work in this sort of environment (biological mechanisms are an existence proof), but in the long run, there will be no need to do so. The Silicon Valley Faire simulation gives the right idea: Systems can be built with no loose molecules, making nanomechanical design much easier. If no molecules are loose inside a machine, then loose molecules can't cause problems there.

"Will Chemical Instability Mess Things Up?"

Chemists perform chemical reactions, which means that they tend to work with reactive, unstable molecules. Many molecules, though, can sit around in peace with their neighbors for millions of years, as is known both from chemical theory and from the study of molecules trapped in ancient rock. Nanomachines can be built from the more stable sorts of structure. The only necessary exception is in molecular assembly, where molecules must react, but even here the reactive molecules need not be turned loose. They can be applied just when and where they are needed in the construction process.

"Is It Too Complex, Like Biology?"

An easy way to explain molecular manufacturing is to say that it is somewhat like molecular biology: small, complex molecular devices working together to build things and do various jobs. The next point, however, is that molecular manufacturing is different in every detail and different in overall structure: compare the nanocomputers, assembler arms, and conveyor belts described above to the shaggy, seething living cell described in the last chapter. Biology is complex in a strange and wonderful way. Engineers need not even *understand* life, much less duplicate it, merely to build a molecular-scale factory.

"I don't see anything wrong with it. But it's so interdisciplinary—couldn't there be a problem I can't see?"

Nanotechnology is basically a shotgun marriage of chemistry and mechanical engineering, with physics (as always) presiding. This makes a complete evaluation difficult for most of today's specialists, because each of these fields is taught separately and usually practiced separately. Many specialists, having highly focused backgrounds, find themselves unequipped to evaluate proposals that overlap other disciplines. When asked to do so, they will state feelings of discomfort, because although they can't identify any particular problems, they can't verify the entire concept as sound. Scientists and engineers with multidisciplinary backgrounds, or with access to specialists from other fields, can evaluate the idea from all sides. We'll meet some of these in Chapter 4.

IT WILL WORK.

When physicists, chemists, biologists, engineers, and computer scientists evaluate those parts of nanotechnology that fall within their disciplines, they agree: At no point would it require new principles or violate a physical law. There may for many years be some experts offering negative off-the-cuff opinions, but the consensus among those who have taken the time to examine the facts is clear. Molecular nanotechnology falls entirely within the realm of the possible.

"IT WOULD WORK, BUT ISN'T IT A BAD IDEA TO IMPLEMENT IT?"

If this means, "These new technologies *could easily* do far more harm than good," then there is no argument, because no one seems to disagree.

If this means, "These new technologies *will certainly* do more harm than good," then we disagree: much good is possible, much harm is avoidable, and it would be too bold to declare any such outcome "certain."

If this means, "These new technologies should be avoided," then we reply, "How, with what risks, and with what consequences?"

Chapters 12 and 13 conclude that it is safer to ride the beast than to hang on to its tail while others swarm aboard.

If this means, "Don't think about it or describe it," then we reply, "How else are we to understand it or make decisions?"

Increased human abilities have routinely been used to damage the environment and to make war. Even the crude technologies of the twentieth century have taken us to the brink. It is natural to feel exhilarated (or terrified) by a prospect that promises (or threatens) to extend human abilities beyond most past dreams (or nightmares). It is better to feel both, to meld and moderate these feelings, and to set out on a course of action that makes bad outcomes less likely. We're convinced that the best course is to focus on the potential good while warning of the potential evils.

"BUT ISN'T IT UNLIKELY TO ARRIVE WITHIN OUR LIFETIMES?"

Those in failing health may be justified in saying this; others are expressing an opinion that may well be wrong. It would be optimistic to assume that benefits are around the corner, and prudent to assume that they will be long delayed. Conversely, it would be optimistic to assume that dangers will be long delayed, and prudent to assume that they will arrive promptly. Whatever good or ill may come of post-breakthrough capabilities, the turbulence of the coming transition will present a real danger. While we invite readers to take a "What if?" stance toward these technologies, it would be imprudent to listen to the lulling sound of the promise "not in our lifetimes."

> Even today, public acceptance of man's coming exploration of space is slow. It is considered an event our children may experience, but certainly not one that we shall see.
> —E. Bergaust and W. Beller
> From the foreword to *Satellite!*, written July 1957
> *Sputnik* orbits Earth, October 1957
> Footprints on Moon, July 1969

PERSPECTIVE

We are still many years away from nanotechnology based on molecular manufacturing. It might even seem that such vast, slow giants as ourselves could never make such small, quick machines. The following sections will describe how advances in science and technology are leading toward these abilities. We'll try to get some feel for the road ahead, for its length, and for how fast we're moving. We are already surprisingly close to developing a crude molecular manufacturing technology, and getting visibly closer every week. The first, crude technology will enable the construction of molecular machines that can be used to build better molecular machines, climbing a ladder of capabilities that leads to general-purpose molecular assemblers as good or better than those described here.

The opportunities then will be enormous. If we haven't prepared, the dangers, too, will be enormous. Whether we're ready or not, the resulting changes will be disruptive, sweeping industries aside, upending military strategies, and transforming our ways of life.

FOUR

Paths, Pioneers, and Progress

Abasic question about nanotechnology is, "When will it be achieved?" The answer is simple: No one knows. How molecular machines will behave is a matter for calculation, but how long it will take us to develop them is a separate issue. Technology timetables can't be calculated from the laws of nature, they can only be guessed at. In this chapter, we examine different paths to nanotechnology, hear what some of the pioneers have to say, and describe the progress already made. This will not answer our basic question, but it will educate our guesses.

Molecular nanotechnology could be developed in any of several basically different ways. Each of these basic alternatives itself includes further alternatives. Researchers will be asking, "How can we make the fastest progress?" To understand the answers they may come to, we need to ask the same question here, adopting (for the moment) a gung-ho, let's-go, how-do-we-get-the-job-done? attitude. We give some of the researchers' answers in their own words.

WILL IT EVER BE ACHIEVED?

Like "When will it be achieved?", this is a basic question with an answer beyond calculation. Here, though, the answer seems fairly clear. Throughout history, people have worked to achieve better control of matter, to convince atoms to do what we want them to do. This has gone on since before people learned that atoms exist, and has accelerated ever since. Although different industries use different materials and different tools and methods, the basic aim is always the same. They seek to make better things, and make them more consistently, and that means better control of the structure of matter. From this perspective, nanotechnology is just the next, natural step in a progression that has been under way for millennia.

Consider the compact discs now replacing older stereo records: Both the old and the new technologies stamp patterns into plastic, but for CDs, the bumps on the stamping surface are only about 130 by 600 nanometers in size, versus 100,000 nanometers or so for the width of the groove on an old-style record. Or look at a personal computer. John Foster, a physicist at IBM's Almaden Research Center, points to a hard disk and says that "inside that box are a bunch of whirring disks, and every one of those disks has got a metal layer where the information is stored. The last thing on top of the metal layer is a monolayer that's the lubricant between the disk and the head that flies over it. The monolayer is not fifteen angstroms [15 angstroms = 1.5 nanometers] and it's not three, because fifteen won't work and neither will three. So it has to be ten plus or minus a few angstroms. This is definitely working in the nanometer regime. We're at that level: We ship it every day and make money on it every day."

The transistors on computer chips are heading down in size on an exponential curve. Foster's colleague at IBM, Patrick Arnett, expects the trend to continue: "If you stay on that curve, then you end up at the atomic scale at 2020 or so. That's the nature of technology now. You expect to follow that curve as far as you can go." The trend is clear, and at least some of the results can be foreseen, but the precise path and timetable for the development of nanotechnology is unpredictable. This unpredictability goes to the heart of important questions: "How will this technology be developed? Who will do it? Where? When? In ten years? Fifty? A hundred? Will this happen in *my* lifetime?" The answers will depend on what people do with their time and resources, which in turn will depend on what goals they think are most promising. Human attitudes, understanding, and goals will make all the difference.

WHAT DECISIONS MOST AFFECT THE RATE OF ADVANCE?

Decisions about research directions are central. Researchers are already pouring effort into chemical synthesis, molecular engineering, and related fields. The same amount of effort could produce more impressive results in molecular nanotechnology if a fraction of it were differently directed. The research funders—corporate executives, and decision makers in science funding agencies like the National Science Foundation in the United States and Japan's Ministry of International Trade and Industry—all have a large influence on research directions, but so do the researchers working in the labs. They submit proposals to potential funders (and often spend time on personally chosen projects, regardless of funding), so their opinions also shape what happens. Where public money is involved, politicians' impressions of public opinion can have a huge influence, and public opinion depends on what all of us think and say.

Still, researchers play a central role. They tend to work on what they think is interesting, which depends on what they think is possible, which depends on the tools they have or—among the most creative researchers—on the tools they can see how to make. Our tools shape how we think: as the saying goes, when all you have is a hammer, everything looks like a nail. New tools encourage new thoughts and enable new achievements, and decisions about tool development will pace advances in nanotechnology. To understand the challenges ahead, we need to take a look at ideas about the tools that will be needed.

WHY ARE TOOLS SO IMPORTANT?

Throughout history, limited tools have limited achievement. Leonardo da Vinci's sixteenth-century chain drives and ball bearings were theoretically workable, yet never worked in their inventor's lifetime. Charles Babbage's nineteenth-century mechanical computer suffered the same fate. The problem? Both inventors needed precisely machined parts that (though readily available today) were beyond the manufacturing technology of their times. Physicist David Miller recounts how a sophisticated integrated circuit design project at TRW hit similar limits in the early 1980s: "It all came down to whether a German company could cool their glass lenses slowly enough to give us the accuracy we needed. They couldn't."

In the molecular world, tool development again paces progress, and new tools can bring breathtaking advances. Mark Pearson, director of molecular biology for Du Pont, has observed this in action: "When I was a graduate student back in the 1950s, it was a multiyear problem to determine the molecular structure of a single protein. We used to say, 'One protein, one career.' Yet now the time has shrunk from a career to a decade to a year—and in optimal cases to a few months." Protein structures can be mapped atom by atom by studying X-ray reflections from layers in protein crystals. Pearson observes that "Characterizing a protein was a career-long endeavor in part because it was so difficult to get crystals, and just getting the material was a big constraint. With new technologies, we can get our hands on the material now—that may sound mundane, but it's a great advance. To the people in the field, it makes all the difference in the world." Improved tools for making and studying proteins are of special importance because proteins are promising building blocks for first-generation molecular machines.

BUT ISN'T SCIENCE ABOUT DISCOVERIES, NOT TOOLS?

Nobel Prizes are more often awarded for discoveries than for the tools (including instruments and techniques) that made them possible. If the goal is to spur scientific progress, this is a shame. This pattern of reward extends throughout science, leading to a chronic underinvestment in developing new tools. Philip Abelson, an editor of the journal *Science*, points out that the United States suffers from "a lack of support for development of new instrumentation. At one time, we had a virtual monopoly in pioneering advances in instrumentation. Now practically no federal funds are available to universities for the purpose." It's easier and less risky to squeeze one more piece of data out of an existing tool than to pioneer the development of a new one, and it takes less imagination.

But new tools emerge anyway, often from sources in other fields. The study of protein crystals, for example, can benefit from new X-ray sources developed by physicists, and techniques from chemistry can help make new proteins. Because they can't anticipate tools resulting from innovations in other fields, scientists and engineers are often too pessimistic about what can be achieved in their own fields. Nanotechnology will join several fields, and yield tools useful in many others. We should expect surprising results.

WHAT TOOLS DO RESEARCHERS USE TO BUILD SMALL DEVICES?

Today's tools for making small-scale structures are of two kinds: molecular-processing tools and bulk-processing tools. For decades, chemists and molecular biologists have been using better and better molecular-processing tools to make and manipulate precise, molecular structures. These tools are of obvious use. Physicists, as we will see, have recently developed tools that can also manipulate molecules. Combined with techniques from chemistry and molecular biology, these physicist's tools promise great advances.

Microtechnologists have applied chip-making techniques to the manufacture of microscopic machines. These technologies—the main approach to miniaturization in recent decades—can play at most a supporting role in the development of nanotechnology. Despite appearances, it seems that microtechnology cannot be refined into nanotechnology.

BUT ISN'T NANOTECHNOLOGY JUST VERY SMALL MICROTECHNOLOGY?

For many years, it was conventional to assume that the road to very small devices led through smaller and smaller devices: a top-down path. On this path, progress is measured by miniaturization: How small a transistor can we build? How small a motor? How thin a line can we draw on the surface of a crystal? Miniaturization focuses on scale and has paid off well, spawning industries ranging from watchmaking to microelectronics.

Researchers at AT&T Bell Labs, the University of California at Berkeley, and other laboratories in the United States have used micromachining (based on microelectronic technologies) to make tiny gears and even electric motors. Micromachining is also being pursued successfully in Japan and Germany. These microgears and micromotors are, however, enormous by nanotechnological standards: a typical device is measured in tens of micrometers, billions of times the volume of comparable nanogears and nanomotors. (In our simulated molecular world, ten microns is the size of a small town.) In size, confusing microtechnology with molecular nanotechnology is like confusing an elephant with a ladybug.

The differences run deeper, though. Microtechnology dumps atoms on surfaces and digs them away again in bulk, with no regard for which atom goes where. Its methods are inherently crude. Molecular nanotechnology, in contrast, positions each atom with care. As Bill DeGrado, a protein chemist at Du Pont, says, "The essence of nanotechnology is that people have worked for years making things smaller and smaller until we're approaching molecular dimensions. At that point, one can't make smaller things except by starting with molecules and building them up into assemblies." The difference is basic: In microtechnology, the challenge is to build smaller; in nanotechnology, the challenge is to build bigger—we can already make *small* molecules.

(A language warning: In recent years, *nanotechnology* has indeed been used to mean "very small microtechnology"; for this usage, the answer to the above question is yes, by definition. This use of a new word for a mere extension of an old technology will produce considerable confusion, particularly in light of the widespread use of *nanotechnology* in the sense found here. Nanolithography, nanoelectronics, nanocomposites, nanofabrication: not all that is *nano-* is molecular, or very relevant to the concerns raised in this book. The terms *molecular nanotechnology* and *molecular manufacturing* are more awkward but avoid this confusion.)

WILL MICROTECHNOLOGY LEAD TO NANOTECHNOLOGY?

Can bulldozers can be used to make wristwatches? At most, they can help to build factories in which watches are made. Though there could be surprises, the relevance of microtechnology to molecular nanotechnology seems similar. Instead, a bottom-up approach is needed to accomplish engineering goals on the molecular scale.

WHAT ARE THE MAIN TOOLS USED FOR MOLECULAR ENGINEERING?

Almost by definition, the path to molecular nanotechnology must lead through molecular engineering. Working in different disciplines, driven by different goals, researchers are making progress in this field. Chemists are developing techniques able to build precise molecular structures of sorts never before seen. Biochemists are learning to build structures of familiar kinds, such as proteins, to make new molecular objects.

In a visible sense, most of the tools used by chemists and biochemists are rather unimpressive. They work on countertops cluttered with dishes, bottles, tubes, and the like, mixing, stirring, heating, and pouring liquids—in biochemistry, the liquid is usually water with a trace of material dissolved in it. Periodically, a bit of liquid is put into a larger machine and a strip of paper comes out with a graph printed on it. As one might guess from this description, research in the molecular sciences is usually much less expensive than research in high-energy physics (with its multibillion-dollar particle accelerators) or research in space (with its multibillion-dollar spacecraft). Chemistry has been called "small science," and not because of the size of the molecules.

Chemists and biochemists advance their field chiefly by developing new molecules that can serve as tools, helping to build or study other molecules. Further advances come from new instrumentation, new ways to examine molecules and determine their structures and behaviors. Yet more advances come from new software tools, new computer-based techniques for predicting how a molecule with a particular structure will behave. Many of these software tools let researchers peer through a screen into simulated molecular worlds much like those toured in the last two chapters.

Of these fields, it is biomolecular science that is most obviously developing tools that can build nanotechnology, because biomolecules already form molecular machines, including devices resembling crude assemblers. This path is easiest to picture, and can surely work, yet there is no guarantee that it will be fastest: research groups following another path may well win. Each of these paths is being pursued worldwide, and on each, progress is accelerating.

Physicists have recently contributed new tools of great promise for molecular engineering. These are the *proximal probes*, including the *scanning tunneling microscope* (STM) and the *atomic force microscope* (AFM). A proximal-probe device places a sharp tip in proximity to a surface and uses it to probe (and sometimes modify) the surface and any molecules that may be stuck to it.

How Does an STM Work?

An STM brings a sharp, electrically conducting needle up to an electrically conducting surface, *almost* touching it. The needle and surface are electrically connected (see the left-hand side of Figure 4),

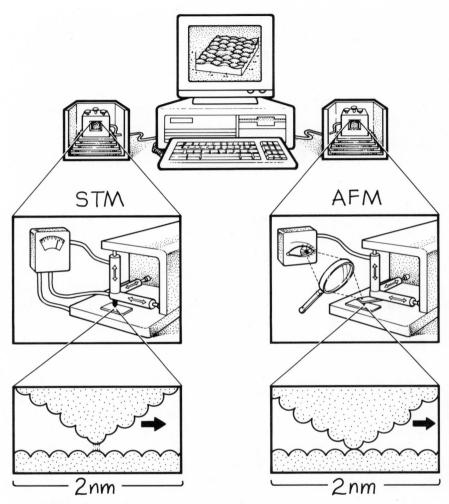

FIGURE 4: STM/AFM

The scanning tunneling microscope (STM, on the left) images surfaces well enough to show individual atoms, sensing surface contours by monitoring the current jumping the gap between tip and surface. The atomic force microscope (AFM, on the right) senses surface contours by mechanical contact, drawing a tip over the surface and optically sensing its motion as it passes over single-atom bumps.

so that a current will flow if they touch, like closing a switch. But at just what point do soft, fuzzy atoms "touch"? It turns out that a detectable current flows when just two atoms are in tenuous contact—fuzzy fringes barely overlapping—one on the surface and one on the tip of the needle. By delicately maneuvering the needle around over the surface, keeping the current flowing at a tiny, constant rate, the STM can map shape of the surface with great precision. Indeed, to keep the current constant, the needle has to go up and down as it passes over individual atoms.

The STM was invented by Gerd Binnig and Heinrich Rohrer, research physicists studying surface phenomena at IBM's research labs in Zurich, Switzerland. After working through the 1970s, Rohrer and Binnig submitted their first patent disclosure on an STM in mid-1979. In 1982, they produced images of a silicon surface, showing individual atoms. Ironically, the importance of their work was not immediately recognized: Rohrer and Binnig's first scientific paper on the new tool was rejected for publication on the grounds that it was "not interesting enough." Today, STM conferences draw interested researchers by the hundreds from around the world.

In 1986—quite promptly as these things go—Binnig and Rohrer were awarded a Nobel Prize. The Swedish Academy explained its reasoning: "The scanning tunneling microscope is completely new and we have so far seen only the beginning of its development. It is, however, clear that entirely new fields are opening up for the study of matter." STMs are no longer exotic: Digital Instruments of Santa Barbara, California, sells its system (the Nanoscope®) by mail with an atomic-resolution-or-your-money-back guarantee. Within three years of their commercial introduction, hundreds of STMs had been purchased.

How Does an AFM Work?

The related atomic force microscope (on the right side of Figure 4) is even simpler in concept: A sharp probe is dragged over the surface, pressed down gently by a straight spring. The instrument senses motions in the spring (usually optically), and the spring moves up and down whenever the tip is dragged over an atom on the surface. The tip "feels" the surface just like a fingertip in the simulated molecular world. The AFM was invented by Binnig, Quate, and Gerber at Stanford University and IBM San Jose in 1985. After the success of the STM, the importance of the AFM was immediately recognized. Among other advantages, it works with nonconducting materials. The next chapter will describe how AFM-based devices might be used as molecular manipulators in developing molecular nanotechnology. As this is written, AFMs have just become commercially available.

(Note that AFMs and STMs are not quite as easy to use as these descriptions might suggest. For example, a bad tip or a bad surface can prevent atomic resolution, and pounding on the table is not recommended when such sensitive instruments are in operation. Further, scientists often have trouble deciding just what they're seeing, even when they get a good image.)

CAN PROXIMAL PROBES MOVE ATOMS?

To those thinking in terms of nanotechnology, STMs immediately looked promising not only for seeing atoms and molecules but for manipulating them. This idea soon became widespread among physicists. As Calvin Quate stated in *Physics Today* in 1986, "Some of us believe that the scanning tunneling microscope will evolve. . . . that one day [it] will be used to write and read patterns of molecular size." This approach was suggested as a path to molecular nanotechnology in *Engines of Creation*, again in 1986.

By now, whole stacks of scientific papers document the use of STM and AFM tips to scratch, melt, erode, indent, and otherwise modify surfaces on a nanometer scale. These operations move atoms around, but with little control. They amount to bulk operations on a tiny scale—*one* fine scratch a few dozen atoms wide, instead of the billions that result from conventional polishing operations.

CAN PROXIMAL PROBES MOVE ATOMS MORE PRECISELY?

In 1987, R. S. Becker, J. A. Golovchenko, and B. S. Swartzen-truber at AT&T Bell Laboratories announced that they had used an STM to deposit small blobs on a germanium surface. Each blob was thought to consist of one or a few germanium atoms. Shortly there-after, IBM Almaden researchers John Foster, Jane Frommer, and Patrick Arnett achieved a milestone in STM-based molecular manip-ulation. Of this team, Foster and Arnett attended the First Foresight Conference on Nanotechnology, where they told us the motivations behind their work.

Foster came to IBM from Stanford University, where he had completed a doctorate in physics and taught at graduate school. The STM work was one of his first projects in the corporate world. He describes his colleague Arnett as a former "semiconductor jock" in-volved in chip creation at IBM's Burlington and Yorktown locations. Besides his doctorate in physics, Arnett brought mechanical-engi-neering training to the effort.

Arnett explains what they were trying to do: "We wanted to see if you could do something on an atomic scale, to create a mechanism for storing information and getting it back reliably." The answer was yes. In January 1988, the journal *Nature* carried their letter reporting success in pinning an organic molecule to a particular location on a surface, using an STM to form a chemical bond by applying an elec-trical pulse through the tip. They found that having created and sensed the feature, they could go back and use another voltage pulse from the tip to change the feature again: enlarging it, partly erasing it, or completely removing it.

IBM quickly saw a commercial use, as explained by Paul M. Horn, acting director of physical sciences at the Thomas J. Watson Research Center: "This means you can create a storage element the size of an atom. Ultimately, the ability to do that could lead to stor-age that is ten million times more dense than anything we have to-day." A broader vision was given by another researcher, J. B. Pethica, in the issue of *Nature* in which the work appeared: "The partial erasure reported by Foster *et al.* implies that molecules may have pieces deliberately removed, and in principle be atomically 'edited,' thereby demonstrating one of the ideals of nanotechnology."

CAN PROXIMAL PROBES MOVE ATOMS WITH COMPLETE PRECISION?

Foster's group succeeded in pinning single molecules to a surface, but they couldn't control the results—the position and orientation—precisely. In April 1990, however, another group at the same laboratory carried the manipulation of atoms even further, bringing a splash of publicity. Admittedly, the story must have been hard to resist: It was accompanied by an STM picture of the name "IBM," spelled out with thirty-five precisely placed atoms (Figure 5). The precision here is complete, like the precision of molecular assembly: each atom sits in a dimple on the surface of a nickel crystal; it can rest either in one dimple or in another, but never somewhere between.

Donald Eigler, the lead author on the *Nature* paper describing this work, sees clearly where all this is leading: "For decades, the electronics industry has been facing the challenge of how to build smaller and smaller structures. For those of us who will now be using individual atoms as building blocks, the challenge will be how to build up structures atom by atom."

HOW FAR CAN PROXIMAL PROBES TAKE US?

Proximal probes have advantages as a tool for developing nano-technology, but also weaknesses. Today, their working tips are rough and irregular, typically even rougher than shown in Figure 5. To make stable bonds form, John Foster's group used a pulse of electricity, but the results proved hard to control. The "IBM" spelled out by Donald Eigler's group was precise, but stable only at temperatures

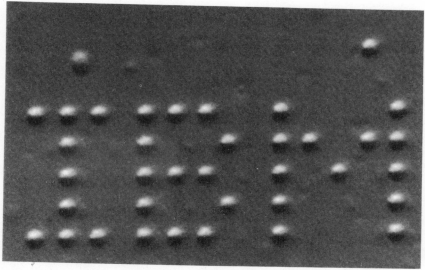

FIGURE 5: WORLD'S SMALLEST LOGO—35 XENON ATOMS
(Courtesy of IBM Research Division)

near absolute zero—such patterns vanish at room temperature because they are not based on stable chemical bonds. Building structures that are both stable and precise is still a challenge. To form stable bonds in precise patterns is the next big challenge.

John Foster says, "We're exploring a concept which we call 'molecular herding,' using the STM to 'herd' molecules the way my Shetland sheep dog would herd sheep. . . . Our ultimate goal with molecular herding is to make one particular molecule move to another particular one, and then essentially force them together. If you could put two molecules that might be small parts of a nanomachine on the surface, then this kind of herding would allow you to haul one of them up to the other. Instead of requiring random motion of a liquid and specific chemical lock-and-key interactions to give you exactly what you want in bringing two molecules together [as in chemical and biochemical approaches], you could drive that reaction on a local level with the STM. You could use the STM to put things where you want them to be." The next chapter will discuss additional ideas for using proximal probes in early nanotechnology.

Proximal-probe instruments may be a big help in building the first generation of nanomachines, but they have a basic limit: Each instrument is huge on a molecular scale, and each could bond only one molecular piece at a time. To make anything large—say, large enough to see with the naked eye—would take an absurdly long time. A device of this sort could add one piece per second, but even a pinhead contains more atoms than the number of seconds since the formation of Earth. Building a Pocket Library this way would be a long-term project.

How Can Such Slow Systems Ever Build Anything Big?

Rabbits and dandelions contain structures put together one molecular piece at a time, yet they grow and reproduce quickly. How? They build in parallel, with many billions of molecular machines working at once. To gain the benefits of such enormous parallelism, researchers can either use proximal probes to build a better, next-generation technology, or use a different approach from the start.

The techniques of chemistry and biomolecular engineering already have enormous parallelism and already build precise molecular structures. Their methods, however, are less direct than the still hypothetical proximal probe–based molecule-positioners. They use molecular building blocks shaped to fit together spontaneously, in a process of *self-assembly*.

David Biegelsen, a physicist who works with STMs at the Xerox Palo Alto Research Center, put it this way at the nanotechnology conference: "Clearly, assembly using STMs and other variants will have to be tried. But biological systems are an existence proof that assembly and self-assembly can be done. I don't see why one should try to deviate from something that already exists."

What Are the Main Advantages of Molecular Building Blocks?

A huge technology base for molecular construction already exists. Tools originally developed by biochemists and biotechnologists to deal with molecular machines found in nature can be redirected to make new molecular machines. The expertise built up by chemists in more than a century of steady progress will be crucial in molecular design and construction. Both disciplines routinely handle molecules by the billions and get them to form patterns by self-assembly. Biochemists, in particular, can begin by copying designs from nature.

Molecular building-block strategies could work together with proximal-probe strategies, or could replace them, jumping directly to the construction of large numbers of molecular machines. Either way, protein molecules are likely to play a central role, as they do in nature.

How Can Protein Engineering Build Molecular Machines?

Proteins can self-assemble into working molecular machines, objects that *do* something, such as cutting and splicing other molecules or making muscles contract. They also join with other molecules to form huge assemblies like the ribosome (about the size of a washing machine, in our simulation view). Ribosomes—programmable machines for manufacturing proteins—are nature's closest approach to a molecular assembler. The genetic-engineering industry is chiefly in the business of reprogramming natural nanomachines, the ribosomes, to make new proteins or to make familiar proteins more cheaply. Designing new proteins is termed *protein engineering*. Since biomolecules already form such complex devices, it's easy to see that advanced protein engineering could be used to build first-generation nanomachines.

IF WE CAN MAKE PROTEINS, WHY AREN'T WE BUILDING FANCY MOLECULAR MACHINES?

Making proteins is easier than designing them. Protein chemists began by studying proteins found in nature, but have only recently moved on to the problem of engineering new ones. These are called *de novo* proteins, meaning completely new, made from scratch. Designing proteins is difficult because of the way they are constructed. As Bill DeGrado, a protein chemist at Du Pont, explains: "A characteristic of proteins is that their activities depend on their three-dimensional structures. These activities may range from hormonal action to a function in digestion or in metabolism. Whatever their function, it's always essential to have a definite three-dimensional shape or structure." This three-dimensional structure forms when a chain folds to form a compact molecular object. To get a feel for how tough it is to predict the natural folding of a protein chain, picture a straight piece of cord with hundreds of magnets and sticky knots along its length. In this state, it's easy to make and easy to understand. Now pick it up, put it in a glass jar, and shake it for a long time. Could you predict its final shape? Certainly not: it's a tangled mess. One might call this effort at prediction "the sticky-cord–folding problem"; protein chemists call theirs "the protein-folding problem."

Given the correct conditions, a protein chain always folds into one special shape, but that shape is hard to predict from just the straightened structure. Protein designers, though, face the different job of first determining a desired final shape, and then figuring out what linear sequence of amino acids to use to make that shape. Without solving the classic protein-folding problem, they have begun to solve the protein-*design* problem.

WHAT HAS BEEN ACCOMPLISHED SO FAR?

Bill DeGrado and his colleagues at Du Pont had one of the first successes: "We've been able to use basic principles to design and build a simple molecule that folds up the way we want it to. This is really the first real example of a designed protein structure, designed from scratch, not by taking an already existing structure and tinkering with it."

Although scientists do the work, the work itself is really a form of engineering, as shown by the title of the field's journal, *Protein Engineering*. Bill DeGrado's description of the process makes this clear: "After you've made it, the next step is to find out whether your protein did what you expected it to do. Did it fold? Did it pass ions across bilayers [such as cell membranes]? Does it have a catalytic function [speeding specific chemical reactions]? And that's tested using the appropriate experiment. More than likely, it won't have done what you wanted it to do, so you have to find out why. Now, a good design has in it a contingency plan for failure and helps you learn from mistakes. Rather than designing a structure that would take a year or more to analyze, you design it so that it can be assayed for given function or structure in a matter of days."

Many groups are pursuing protein design today, including academic researchers like Jane and Dave Richardson at Duke University, Bruce Erickson at the University of North Carolina, and Tom Blundell, Robin Leatherbarrow, and Alan Fersht in Britain. The successes have started to roll in. Japan, however, is unique in having an organization devoted exclusively to such projects: the Protein Engineering Research Institute (PERI) in Osaka. In 1990, PERI announced the successful design and construction of a *de novo* protein several times larger than any built before.

Is There Anything Special About Proteins?

The main advantage of proteins is that they are familiar: a lot is known about them, and many tools exist for working with them. Yet proteins have disadvantages as well. Just because this design work is starting with proteins—soft, squishy molecules that are only marginally suitable for nanotechnology—doesn't mean it will stay within those limits. DeGrado points out, "The fundamental goal of our work in *de novo* design is to be able to take the next step and get entirely away from protein systems." An early example is the work of Wallace Carothers of Du Pont, who used a *de novo* approach to studying the nature of proteins: Rather than trying to cut up proteins, he tried to build up things starting with amino acids and other similar monomers. In 1935, he succeeded in making nylon.

DeGrado explains; "There is a deep philosophical belief at Du Pont in the ability of people to make molecules *de novo* that will do useful things. And there is a fair degree of commitment from the management that following that path will lead to products: not directly, and not always predictably, but they know that they need to support the basic science.

"I think ultimately we have a better chance at doing some really exciting things by *de novo* design, because our repertory should be much greater than that of nature. Think about the ability to fly: One could breed better carrier pigeons or one could design airplanes." The biology community, however, leans more toward ornithology than toward aerospace engineering. DeGrado's experience is that "a lot of biologists feel that if you aren't working with the real thing [natural proteins], you aren't studying biology, so they don't totally accept what we're doing. On the other hand, they recognize it as good chemistry."

WHERE IS PROTEIN ENGINEERING HEADED?

Like the IBM physicists, protein designers are moved by a vision of molecular engineering. In 1989, Bill DeGrado predicted, "I think we'll be able to make catalysts or enzymelike molecules, possibly ones that catalyze reactions not catalyzed in nature." Catalysts are molecular machines that speed up chemical reactions: they form a shape for the two reacting molecules to fit into and thereby help the reaction move faster, up to a million reactions per second. New ones, for reactions that now go slowly, will give enormous cost savings to the chemical industry.

This prediction was borne out just a few months later, when Denver researchers John Stewart, Karl Hahn, and Wieslaw Klis announced their new enzyme, designed from scratch over a period of two years and built successfully on the first try. It's a catalyst, making some reactions go about 100,000 times faster. Nobel Prize-winning biochemist Bruce Merrifield believes that "if others can reproduce and expand on this work, it will be one of the most important achievements in biology or chemistry."

DeGrado also has longer-term plans for protein design, beyond making new catalysts: "It will allow us to think about designing molecular devices in the next five to ten years. It should be possible ultimately to specify a particular design and build it. Then you'll have, say, proteinlike molecules that self-assemble into complex molecular objects, which can serve as machinery. But there's a limit to how small you can make devices. You'll shrink things down so far and then you won't be able to go any further, because you've reached molecular dimensions."

Mark Pearson shows that management at Du Pont also has this vision. Regarding the prospects for nanotechnology and assemblers, he remarked, "You know, it'll take money and effort and good ideas for sure. But to my way of thinking, there is no absolute fundamental limitation to preclude us from doing this kind of thing." He didn't say his company plans to develop nanotechnology, but such plans aren't really necessary. Du Pont is already on the nanotechnology path, for other—shorter-term, commercial—reasons. Like IBM, if they do decide to move quickly, they have the resources and forward-looking people needed to succeed.

WHO ELSE BUILDS MOLECULAR OBJECTS?

Chemists, most of whom do *not* work on proteins, are the traditional experts in building molecular objects. As a group, they've been building molecules for over a century, with ever-increasing ability and confidence. Their methods are all indirect: they work with billions of atoms at a time—massive parallelism—but without control of the positions of their workpieces. The molecules typically tumble randomly in a liquid or gas, like pieces of a puzzle that may or may not fit together correctly when shaken together in a box. With clever design and planning, most pieces will join properly.

Chemists mix molecules on a huge scale (in our simulation view, a test tube holds a churning molecular swarm with the volume of an inland sea), yet they still achieve precise molecular transformations. Given that they work so indirectly, their achievements are astounding. This is, in part, the result of the enormous amount of work poured into the field for many decades. Thousands of chemists are working on molecular construction in the United States alone; add to that the chemists in Europe, in Japan, and in the rest of the world, and you have a huge community of researchers making great strides. Though it publishes only a one-paragraph summary of each research report, a guide to the chemical literature—*Chemical Abstracts*—covers several library walls and grows by many feet of shelf space every year.

HOW CAN MIXING CHEMICALS BUILD MOLECULAR OBJECTS?

An engineer would say that chemists (at least those specializing in synthesis) are doing construction work, and would be amazed that they can accomplish anything without being able to grab parts and put them in place. Chemists, in effect, work with their hands tied behind their backs. Molecular manufacturing can be termed "positional chemistry" or "positional synthesis," and will give chemists the ability to put molecules where they want them in three-dimensional space. Rather than trying to design puzzle pieces that will stick together properly by themselves when shaken together in a box, chemists will then be able to treat molecules more like bricks to be stacked. The basic principles of chemistry will be the same, but strategies for construction will become far simpler.

Without positional control, chemists face a problem something like this: Picture a giant glass barrel full of tiny battery-powered drills, buzzing away in all directions, vibrating around in the barrel. Your goal is to take a piece of wood and put a hole in just one specific spot. If you simply throw it in the barrel, it will be drilled haphazardly in many places. To control the process, you must protect all the places you don't want drilled—perhaps by gluing protective pieces of metal over most of the wood surface. This problem—how to protect one part of a molecule while altering another part—has forced chemists to develop ever-cleverer ploys to build larger and larger molecules.

If Chemists Can Make Molecules, Why Aren't They Building Fancy Molecular Machines?

Chemists can achieve great things, but have focused much of their effort on duplicating molecules found in nature and then making minor variants. As an example, take palytoxin, a molecule found in a Hawaiian coral. It was so difficult to make in the lab that it has been called "the Mount Everest of synthetic chemistry," and its synthesis was hailed as a triumph. Other efforts are poured into making small molecules with unusual bonding, or molecules of remarkable symmetry, like "cubane" and "dodecahedrane" (shaped like the Platonic solids they are named after).

Chemists, at least in the United States, regard themselves as natural scientists even when their life's work is the construction of molecules by artificial means. Ordinarily, people who build things are called engineers. And indeed, at the University of Tokyo the Department of Synthetic Chemistry is part of the Faculty of Engineering; its chemists are designing molecular switches for storing computer data. Engineering achievements will require work directed at engineering goals.

HOW COULD CHEMISTS MOVE TOWARD BUILDING MOLECULAR MACHINES?

Molecular engineers working toward nanotechnology need a set of molecular building blocks for making large, complex structures. Systematic building-block construction was pioneered by Bruce Merrifield, winner of the 1984 Nobel Prize in Chemistry. His approach, known as "solid phase synthesis," or simply "the Merrifield method," is used to synthesize the long chains of amino acids that form proteins. In the Merrifield method, cycles of chemical reaction each add one molecular building block to the end of a chain anchored to a solid support. This happens in parallel to each of trillions of identical chains, building up trillions of molecular objects with a particular sequence of building blocks. Chemists routinely use the Merrifield method to make molecules larger than palytoxin, and related techniques are used for making DNA in so-called gene machines: an ad from an Alabama company reads, "Custom DNA—Purified and Delivered in 48 hours."

While it's hard to predict how a natural protein chain will fold—they weren't designed to fold predictably—chemists could make building blocks that are larger, more diverse, and more inclined to fold up in a single, obvious, stable pattern. With a set of building blocks like these, and the Merrifield method to string them together, molecular engineers could design and build molecular machines with greater ease.

HOW DO RESEARCHERS DESIGN WHAT THEY CAN'T SEE?

To make a new molecule, both its structure and the procedure to make it must be designed. Compared to gigantic science projects like the Superconducting Supercollider and the Hubble Space Telescope, working with molecules can be done on a shoestring budget. Still, the costs of trying many different procedures add up. To help predict in advance what will work and what won't, designers turn to models.

You may have played with molecular models in chemistry class: colored plastic balls and sticks that fit together like Tinker Toys. Each color represents a different kind of atom: carbon, hydrogen, and so on. Even simple plastic models can give you a feel for how many bonds each kind of atom makes, how long the bonds are, and at what angles they are made. A more sophisticated form of model uses only spheres and partial spheres, without sticks. These colorful, bumpy shapes are called CPK models, and are widely used by professional chemists. Nobel laureate Donald Cram remarks that "We have spent hundreds of hours building CPK models of potential complexes and grading them for desirability as research targets." His research, like that of fellow Nobelists Charles J. Pedersen and Jean-Marie Lehn, has focused on designing and making medium-sized molecules that self-assemble.

Although physical models can't give a good description of how molecules bend and move, computer-based molecules can. Computer-based modeling is already playing a key role in molecular engineering. As John Walker (a founder and leader of Autodesk) has remarked, "Unlike all of the industrial revolutions that preceded it, molecular engineering requires, as an essential component, the ability to design, model, and simulate molecular structures using computers."

This has not gone unnoticed in the business community. John Walker's remark was part of a talk on nanotechnology given at Autodesk, a leader in computer-aided design and one of the five largest software firms in the United States. Soon after this talk, the company made its first major investment in the computer-aided design of molecules.

How Does Molecular Design Compare to More Familiar Kinds of Engineering?

Manufacturers and architects know that designs for new products and buildings are best done on a computer, by computer-aided design (CAD). The new molecular-design software can be called *molecular CAD*, and in its forefront are researchers such as Jay Ponder of the Yale University Department of Molecular Biophysics and Biochemistry. Ponder explains that "There's a strong link between what molecular designers are doing and what architects do. Michael Ward of Du Pont is designing a set of building blocks for a Tinker Toy set so that you can build larger structures. That's exactly what we're doing with molecular modeling techniques.

"All the design and mechanical engineering principles that apply to building a skyscraper or a bridge apply to molecular architecture as well. If you're building a bridge, you're going to model it and see how many trucks can be on the bridge at the same time without it collapsing, what kind of forces you're going to apply to it, whether it can stand up to an earthquake.

"And the same process goes on in molecular design: You're designing pieces and then analyzing the stresses and forces and how they will change and perturb the structure. It's exactly the same as designing and building a building, or analyzing the stresses on any macro-scale structure. I think it's important to get people to think in those terms.

"The molecular designer has to be creative in the same way that an architect has to be creative in designing a building. When people are looking at the interior of a protein structure and trying to redesign it to create a space that will have a particular function, such as binding to particular molecules, that's like designing a room to use as a dining room—one that will fit certain sizes of tables and certain numbers of guests. It's the same thing in both cases: You have to design a space for a function."

Ponder combines chemistry and computer science with an overall engineering approach: "I'm kind of a hybrid. I spend about half my time doing experiments and about half my time writing computer programs and doing computational work. In the laboratory, I create or design molecules to test some of the computational ideas. So I'm at the interface." The engineering perspective helps in thinking about where molecular research can lead: "Even though with nanotechnology we're at the nanometer scale, the structures are still big enough that an awful lot of things are classical. Again, it's really like building bridges—very small bridges. And so there are many almost standard mechanical-engineering techniques for architecture and building structures, such as stress analysis, that apply."

DOESN'T ENGINEERING REQUIRE MORE TEAMWORK THAN SCIENCE DOES?

Getting to nanotechnology will require the work of experts in differing fields: chemists, who are learning how to make molecular machines; computer scientists, who are building the needed design tools; and perhaps STM and AFM experts, who can provide tools for molecular positioning. To make progress, however, these experts must do more than just work, they must work together. Because nanotechnology is inherently interdisciplinary, countries that draw hard lines between their academic disciplines, as the United States does, will find that their researchers have difficulty communicating and cooperating.

In chemistry today, a half-dozen researchers aided by a few tens of students and technicians is considered a large team. In aerospace engineering, enormous tasks like reaching the Moon or building a new airliner are broken down into tasks that are within the reach of small teams. All these small teams work together, forming a large team that may consist of thousands of engineers aided by many thousands of technicians. If chemistry is to move in the direction of molecular-systems engineering, chemists will need to take at least a few steps in this direction.

In engineering, everyone knows that designing a rocket will require skills from many disciplines. Some engineers know structures, others know pumps, combustion, electronics, software, aerodynamics, control theory, and so on and so forth down a long list of disciplines. Engineering managers know how to bring different disciplines together to build systems.

In academic science, interdisciplinary work is productive and praised, but is relatively rare. Scientists don't need to cooperate to have their results fit together: they are all describing different parts of the same thing—nature—so in the long run, their results tend to come together into a single picture. Engineering, however, is different. Because it is more creative (it actually *creates* complex things), it demands more attention to teamwork. If the finished parts are going to work together, they must be developed by groups that share a common picture of what each part must accomplish. Engineers in different disciplines are forced to communicate; the challenge of management and team-building is to make that communication happen. This will apply to engineering molecular systems as much as it does to engineering computers, cars, aircraft, or factories.

Jay Ponder suggests that it's a question of perspective. "It's all a matter of what's perceived to be important by the different groups that have to come together to make this work: the chemists doing their bit and the computational people doing their bit. People have to come together and see the big picture. There are people who try to bridge the gaps, but they are rare compared to the people who just work in their own specialty." Progress toward nanotechnology will continue, and as it does, researchers trained as chemists, physicists, and the like will learn to talk to one another to solve new problems. They will either learn to think like engineers and work in teams, or they will be eclipsed by colleagues who do.

ARE THESE PROBLEMS PREVENTING ADVANCES?

With all these problems, the advance toward nanotechnology steadily continues. Industry must gain ever-better control of matter to stay competitive in the world marketplace. The STM, protein engineering, and much of chemistry are driven by commercial imperatives. Focused efforts would yield faster advances, yet even without a clear focus, advances in this direction have an air of inevitability. As Bill DeGrado observes, "We really do have the tools. Experience has shown that when you have the analytic and synthetic tools to do things, in the end science goes ahead and does them—because they are doable." Jay Ponder agrees: "Over the next few years, you're going to see slow evolutionary advances coming from people tinkering with molecular structures and figuring out their principles. People are going to work on a particular problem because they see some application for it or because they got grant funding for it. And in the process of doing something like improving a laundry detergent's ability to clean protein stains, Procter and Gamble is going to help work out the principles for how to increase molecular stability, and to design spaces inside the molecules."

ARE THE JAPANESE BEARING THEIR SHARE OF THE BURDEN IN NANOTECHNOLOGY RESEARCH?

For a variety of reasons, Japan's contribution to nanotechnology research promises to be excellent. While the United States has generally pursued research in this area with little sense of long-term direction, it appears that Japan has begun to take a more focused approach. Researchers there already have clear ideas about molecular machines—about what might work and what probably won't. Japanese researchers are accustomed to a higher level of interdisciplinary contact and engineering emphasis than are Americans. In the United States, we prize "basic science," often calling it "pure science," as if to imply that practical applications are a form of impurity. Japan instead emphasizes "basic technology."

Nanotechnology is a basic technology, and the Japanese recognize it as such. Recent changes at the Tokyo Institute of Technology—Japan's equivalent of MIT—reflect their views of promising directions for future research. For many decades, Tokyo Tech has had two major divisions: a Faculty of Science and a Faculty of Engineering. To these is now being added a Faculty of Bioscience and Biotechnology, to consist of four departments: a Department of Bioscience, a Department of Bioengineering, a Department of Biomolecular Engineering, and what is termed a "Department of Biostructure." The creation of a new faculty in a major Japanese university is a rare event. What U.S. university has a department explicitly devoted to molecular engineering? Japan has both the departments at Tokyo Tech and Kyoto University's recently established Department of Molecular Engineering.

Japan's Institute for Physical and Chemical Research (RIKEN) has broad-based interdisciplinary strength. Hiroyuki Sasabe, head of the Frontier Materials Research Program at RIKEN, notes that the institute has expertise in organic synthesis, protein engineering, and STM technology. Sasabe says that his laboratory may need a molecular manipulator of the sort described in the next chapter to accomplish its goals in molecular engineering.

Research consortia in Japan are also moving toward nanotechnology. The Exploratory Research for Advanced Technology Organization (ERATO) sponsors many three-to-five-year projects in parallel, each with a specific goal. Consider the work in progress:

- Yoshida Nanomechanism Project
- Hotani Molecular Dynamic Assembly Project
- Kunitake Molecular Architecture Project
- Nagayama Protein Array Project
- Aono Atomcraft Project

These focus on different aspects of gaining control over matter at the atomic level. The Nagayama Protein Array Project aims to use proteins as engineering materials to move toward making new molecular devices. The Aono Atomcraft Project does not involve nuclear power—as its translation might imply—but is instead an interdisciplinary effort to use an STM to arrange matter on the atomic scale.

At some point, work on nanotechnology must move beyond spin-offs from other fields and undertake the design and construction of molecular machinery. This shift from opportunistic science to organized engineering requires a change in attitude. In this, Japan leads the United States.

WHAT IS A GOOD EDUCATED GUESS OF HOW LONG IT WILL TAKE TO DEVELOP MOLECULAR NANOTECHNOLOGY?

Molecular nanotechnology will emerge step by step. Major milestones, such as the engineering of proteins and the positioning of individual atoms, have already been passed. To get a sense of the likely pace of developments, we need to look at how various trends fit together.

Computer-based molecular-modeling tools are spawning computer-aided design tools. These will grow more capable. The underlying technology base—computer hardware—has for decades been improving in price and performance on a steeply rising curve, which is generally expected to continue for many years. These advances are quite independent of progress in molecular engineering, but they make molecular engineering easier, speeding advances. Computer models of molecular machines are beginning to appear, and these will whet the appetites of researchers.

Progress in engineering molecular machines, whether using proximal probes or self-assembly, will eventually achieve striking successes; the objectives of research in Japan will begin to draw serious attention; understanding of the long-term promise of molecular engineering will become more widespread. Some combination of these developments will eventually lead to a serious, public appraisal of what these technologies can achieve—and then the world of opinion, funding, and research fashion will change. Before, advances will be steady but haphazard; afterward, advances will be driven with the energy that flows into major commercial, military, and medical research programs, because nanotechnology will be recognized as furthering major commercial, military, and medical goals. The timing of subsequent events depends largely on when this threshold of serious attention is reached.

In making time estimates, people are prone to assume that a large change must take a long time. Most do, but not all. Pocket calculators had a dramatic effect on the slide-rule industry: they replaced it. The speed of this change caught the slide-rule moguls by surprise, but the pace of progress in electronics didn't slow down merely to suit their expectations.

One can argue that nanotechnology will be developed fast: many countries and companies will be competing to get there first. They will be driven onward both by the immense expected benefits—in many areas, including medicine and the environment—as well as by potential military applications. That is a powerful combination of motives, and competition is a powerful accelerator.

A counterargument, though, suggests that development will be slow: anyone who has done anything of significance in the real world of technology—doing a scientific experiment, writing a computer program, bringing a new product to market—knows that these goals take longer than expected. Indeed, Hofstadter's Law states that projects take longer than expected, even when Hofstadter's Law is taken into account. This principle is a good guide for the short term, and for a single project.

The situation differs, though, when many different approaches are being explored by many different groups over a period of years. Most projects may take longer than expected, but with many teams trying many approaches, one approach may prove faster than expected. The winner of a race is always faster than the average runner. John Walker notes, "The remarkable thing about molecular engineering is that it looks like there are many different ways to get there and, at the moment, rapid progress is being made along every path— all at the same time."

Also, technology development is like a race run over an un-

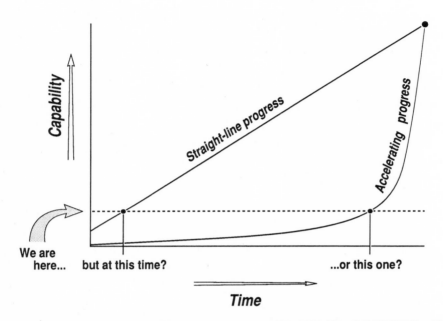

GRAPH OF LINEAR VS. ACCELERATING GROWTH OF TECHNOLOGY K. ERIC DREXLER

How close we are to a goal depends on whether technological advances are at a constant pace or accelerating. In this diagram, the dashed line represents the current level of technology, and the large dot in the upper right represents a goal such as nanotechnology. With a straight-line advance, it's easier to estimate how far away a goal is. With an accelerating advance, a goal can be reached with little warning.

mapped course. When the first runners reach the top of a hill, they may see a shortcut. A trailing runner may decide to crash off into the bushes, and stumble across a bicycle and a paved road. The progress of technology is seldom predictable because progress often reveals new directions.

So how can we estimate a date for the arrival of nanotechnology? It's safest to take a cautious approach: When anticipating benefits, assume it's far off; when preparing for potential problems, assume it's right around the corner. The old folk saying applies: Hope for the best, prepare for the worst. Any dates assigned to "far off" and "right around the corner" can be no better than educated guesses—molecular behavior can be calculated, but not technology timetables of this sort. With those caveats, we would estimate that general-purpose molecular assemblers will likely be developed in the early decades of the twenty-first century, perhaps in the first.

John Walker, whose technological foresight has led Autodesk from start-up to a dominant role in its industry, points out that not long ago, "Many visionaries intimately familiar with the developments of silicon technology still forecast it would take between twenty and fifty years before molecular engineering became a reality. This is well beyond the planning horizon of most companies. But recently, everything has begun to change." Based on the new developments, Walker places his bet: "Current progress suggests the revolution may happen within this decade, perhaps starting within five years."

FIVE

The Threshold
of Nanotechnology

I n the last chapter, we looked at the state of current research, but from there to the nanotechnology of even the Pocket Library scenario is a leap. How will this gap be crossed?

In this chapter, we outline how emerging technologies can lead to nanotechnology. The actual path to nanotechnology—the one that history books will record—could emerge from any one of the research directions in physics, biochemistry, and chemistry recounted in the last chapter, or (more likely) from a combination of them. The availability of so many good options builds confidence that the goal can be reached, even while it decreases confidence that some particular path will be fastest. To see how advances might cross the gap from present technology to early nanotechnology, let's follow one path out of the many possible.

BRIDGING THE GAP

One way to bridge the gap would be through the development of an AFM-based *molecular manipulator* capable of doing primitive molecular manufacturing. This device would combine a simple molecular device—a molecular gripper—with an AFM positioning mechanism. An AFM can move its tip with precision; a molecular manipulator would add a gripper to the tip to hold a molecular tool. A molecular manipulator of this kind would guide chemical reactions by positioning molecules, like a slow, simple, but enormous assembler. (In our standard simulation view, where a molecular-assembler arm fits in a room, the AFM apparatus of a molecular manipulator would be the size of a moon.) Despite its limits, an AFM molecular manipulator will be a striking advance.

How might this advance occur? Since we're choosing one path out of many possible, we may as well include more details and tell a story. (A more technical description of a device like the following can be found in *Nature*; see the Technical Bibliography.)

SCENARIO: DEVELOPING A MOLECULAR MANIPULATOR

Several years ago, researchers at the University of Brobdingnag began work on developing a molecular manipulator. To reach this goal, a team of a dozen physicists, chemists, and protein researchers banded together (some working full time, some part time) and began the creative teamwork needed to solve the basic problems.

First they needed to attach a gripper to an AFM tip. As grippers, they chose fragments of antibody molecules, the selectively sticky proteins that the immune system uses to bind and identify germs. If they could get the "back" of the molecule stuck onto a tip, then the "front" could bind and hold molecular tools. (The advantage of antibody fragments was this: freedom of tool choice.

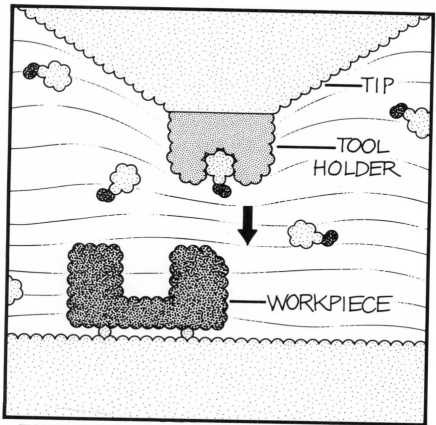

FIGURE 6: MOLECULAR MANIPULATOR

A molecular manipulator (AFM tip and tool holder, above) would bind and position reactive molecular tools to build up a workpiece, molecule by molecule.

Since the late 1980s, researchers had been able to generate antibodies able to bind almost any preselected molecule—or molecular tool.) They tried half a dozen methods before finding one that worked reliably, with results like those shown in Figure 6. A graduate student got her Ph.D., and the AFM tip got its gripper.

In parallel, the U. Brob AFM researchers worked on placing tips in a precise location and then holding them there with atomic accuracy for seconds at a time. This proved straightforward. They used techniques developed elsewhere during the early 1990s, adding only modest refinements.

They now had their gripper and a way of putting it where they wanted it, but they needed a set of tools. The gripper was like the chuck of a drill, waiting to have different bits fitted into its tool-holder slot. So as the final step, the synthetic chemists on the team made a dozen different molecular tools, all identical at one end but different at the other. The similar parts all bound to the same anti-body tool-holder, slotting neatly into position. The different parts were all chemically reactive in different ways. Like the molecular tools in the hall of assembler arms in Chapter 3, each of these tools could use a chemical reaction to transfer some atoms to a molecu-lar object under construction.

Developing the molecular tool kit was the toughest part of the project; it took about as much work as had gone into duplicating the palytoxin molecule back in the 1980s. None of the tasks in the project demanded the solution of a deep scientific puzzle, and none demanded the solution of a notoriously difficult engineering prob-lem. Each task had many possible solutions, the problem was to find a compatible set of solutions and apply them. After a few years, the solutions came together and the U. Brob research team began building new molecules by molecular manipulation. Now many teams are doing likewise.

BUILDING WITH MOLECULAR GRIPPERS AND TOOLS

To build something with the U. Brob team's AFM-based molecu-lar-manipulator system, you use it as follows: First, choose a surface to build on and place it under the tip in a pool of liquid. Then dunk the AFM tip into the liquid, bringing it down to the surface, and back it off a little. Construction can now begin as soon as a tool is loaded into the gripper.

Tubes and pumps can flow different liquids over the surface and past the gripper, carrying different tool molecules. If you want to do something with a tool of Type A, you wash in the proper liquid, and a Type A molecule promptly sticks to the gripper as shown in Figure 6. Once it is in the gripper, you can use the AFM mecha-nism to move it around and put it where you want it. Move it up to the surface at a convenient spot, wait a few seconds, and it reacts, forming a bond and leaving a molecular fragment attached to the spot you chose. To add a different fragment, you can use a tool of Type B: you back up the tip, flow in a fresh liquid carrying the new tools, and in a moment a tool of the new type is bound in place and ready to apply, either on or alongside the first spot. Step by step, you build up a precise molecular structure.

Each step takes only seconds. Molecular tools pop into the gripper in a fraction of a second, and used tools pop off at the same rate. Once the tip has positioned a molecule, it reacts quickly, about a million times faster than unwanted reactions at other sites. In this way, the molecular manipulator gives good control of where reactions will occur (though it is not as reliable as an advanced assembler would be). It is fairly fast by a chemist's standards—per cycle—but still a million times slower than an advanced assembler. It can perform a variety of steps, but isn't as flexible and capable as an advanced assembler. In short, it is hardly the last word in nanotechnology, yet is a great advance over what has gone before.

PRODUCTS

With its ability to accelerate desired reactions by a factor of a million or so, the U. Brob team's molecular manipulator can perform 10,000 to 100,000 steps with good reliability. Back in the 1980s, chemists making protein molecules struggled to perform just one hundred steps. The U. Brob research team (and its many imitators) can now build structures that are stronger and easier to design than proteins: not floppy, folded chains, but rugged objects held together by a sturdy network of bonds. Though not as strong and dense as diamond, these structures are like bits of a tough engineering plastic. A specially adapted computer-aided design system makes it easy to design molecular objects made from these materials.

Yet the AFM-based molecular manipulator has one grave disadvantage: It does chemistry one molecule at a time, and it ties up a machine as expensive as a car for hours or days to produce that one large molecule. Some molecules, though, are valuable enough to be worth building even one at a time. These draw prompt attention.

A single molecule isn't much use as a dye, a drug, or a floor wax, but it can have substantial value if it provides useful information. The U. Brob team quickly publishes a pile of scientific papers based on experiments with single molecules: they build a molecule, probe it, report the results, and build another. Some of these results show chemists elsewhere in the multibillion-dollar chemical industry how to design new catalysts, molecules that can help make other molecules more cheaply, cleanly, and efficiently. This information is worth
a lot.

Three new products of special interest are among the first to be made. The first—molecular electronics—begins with experiments conducted by a research group at a computer-chip company. The team uses their molecular manipulator to build single molecules and probe them, gradually learning how to build the parts needed for molecular electronic computers. These new computers don't immediately become practical, because the costs are too high for making such large molecules with AFM-based technology. Yet some companies begin to produce simpler molecular electronic devices for use in sensors and specialized high-speed signal processing. A specialty industry is born and begins to expand.

The second product is a gene reader, a complex molecular device built on the surface of a chip. The biologists who built the reader combined proteins borrowed from cells with special-purpose molecular machines designed from scratch. The result was a molecular system that binds DNA molecules and pulls them past a read-head-like tape through a tape recorder. The device works as fast as some naturally occurring molecular machines that read DNA, with one key advantage: it outputs its data electronically. At that speed, a single device can read a human genome in about a year. Though still too expensive for a doctor's office, these readers are promptly in great demand from research laboratories. Another small industry is born.

The third product is far more important, in the long run: replacement tips for molecular manipulators, grippers, and tools that are better than the originals. With these new, more versatile devices, researchers are now building more ambitious products and tools.

MORE SCENARIO: THE NEXT STEP TO NANOTECHNOLOGY

While the physicist-led team at U. Brob was finishing its work on the AFM-based molecular manipulator, a chemist-led team at the University of Lilliput was working furiously. They saw the U. Brob desktop machine as too large and its expected products as too expensive. Even back in the 1980s, David Biegelsen of the Xerox Palo Alto Research Center had noted, "The main drawback I see to using a hybrid protoassembler [AFM-based molecular manipulator] is that it would take a long time to build just one unit. Building requires a series of atom-by-atom construction steps. It would be better to build in parallel from the very beginning, making many trillions of these molecules all at the same time. I think there is tremendous power in parallel assembly. Maybe another field, chemistry or biology, offers a better way to do it." The chemists at U. Lill aimed to develop that better way, building first simple and then more and more complex molecular machines. The eventual result was a primitive molecular assembler able to build molecular objects by the trillions.

CHEMISTS' TOOLS

How did the chemists achieve this? During the years when the U. Brob team was developing the molecular manipulator, researchers working in protein science and synthetic chemistry had made better and better systems of molecular building blocks. Chemists were well prepared for doing this: by the late 1980s, it had become possible to build stable structures the size of medium-sized protein molecules, and work had begun to focus on making these molecules perform useful work by binding and modifying other molecules. Chemists learned to use these sophisticated catalysts—early molecular devices—to make their own work easier by helping in the manufacture of still more large molecules.

Another traditional chemist's tool was software for doing computer-aided design. The early software designed by Jay Ponder and Frederic Richards of Yale University ultimately led to semiautomatic tools for designing molecules of a particular shape and function. Chemists then could easily design molecules that would self-assemble into larger structures, several tens of nanometers across.

MOLECULAR CONSTRUCTION MACHINES

These advances in software and chemical synthesis let the U. Lill team tackle the task of building a primitive version of a molecular assembler. Although they couldn't build anything as complex as a nanocomputer or as stiff as diamond, they didn't need to. Their design used sliding molecular rods to position a molecular gripper much like the gripper used at U. Brob, again using the surrounding liquid to control which tool the gripper held. Instead of an AFM's electronic controls, they used the surrounding liquid to control the position of the rods as well. In a neutral solution, the rods would withdraw; in an acid solution, they would extend. How far they moved depended on what other molecules were around to lodge in special pockets and block the motion.

Their primitive assemblers built much the same sorts of products that the U. Brob molecular manipulator did; the tools were similar, and speed and accuracy were about the same. Yet there was one dramatic advantage: About 1,000,000,000,000,000,000,000 U. Lill assemblers could fit in the space occupied by one U. Brob manipulator, and it was easy to produce a mere 1,000,000,000,000,000 times as much product at the same cost.

With the first, primitive assemblers, construction was slow because each step required new liquid baths and several seconds of soaking and waiting, and a typical product might take thousands of steps. Nonetheless, the U. Lill team made a lot of money licensing their technology to researchers trying to commercialize products they had first researched with the U. Brob machine. After starting an independent company (Nanofabricators, Inc.), they poured their

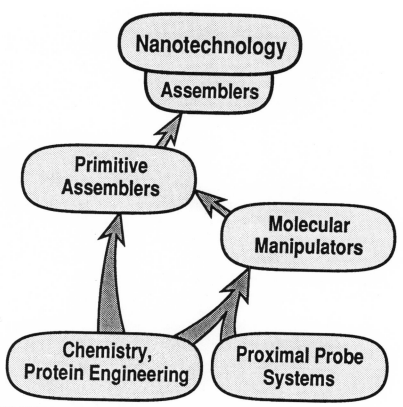

FIGURE 7: PATHS TO NANOTECHNOLOGY K. ERIC DREXLER

research efforts into building better machines. Within a few years, they had assemblers with multiple grippers, each loaded with a different kind of tool; flashes of colored light would flip molecules from state to state (they copied these molecules from the pigments of the retina of the eye); flipping molecules would change tools and change rod positions. Soaking and waiting became a thing of the past, and soon they were pouring out parts that, when mixed with liquid and added to dishes with special blank chips, would build up the dense memory layers that made possible the Pocket Library.

That was when things started moving fast. The semiconductor industry went the way of the vacuum-tube industry. Money and talent poured into the new technology. Molecular CAD tools got better, assemblers made it easy to build what was designed, and fast production and testing made molecular engineering as easy as playing with software. Assemblers got better, faster, and cheaper. Researchers used assemblers to build nanocomputers, and nanocomputers to control better, faster assemblers. Using tools to build better tools is an ancient story. Within a decade, almost anything could be made by molecular manufacturing, and was.

Will developments in the late pre-breakthrough days be as just described? Certainly not: the technical approaches will differ, and the U.S. academic research setting implied by the scenario could easily be replaced by academic, commercial, governmental, or military research settings in any of the advanced nations. What do seem realistic are the implied requirements for effort, technology, and time, as well as the basic capabilities of different devices. We are approaching a threshold of capability beyond which further advances will become easy and fast.

SIX

Working with Nanotechnology

The word *manufacturing* comes from the Latin *manufactus*, meaning "handmade." Today, the term brings to mind huge, noisy machines stamping out products and spewing waste. Giving up manufactured products isn't popular or practical—almost everything we use today is manufactured. If all machine-made products were to suddenly vanish, most people in the United States would find themselves naked and outdoors, with very little around them. Expanding manufacturing is an object of nearly every nation on Earth.

We can't give up manufacturing, but we can replace today's technologies with something radically different. Molecular manufacturing can help us get what we seem to want: high-quality products made at low cost with little environmental impact. Chapter 12 will describe the grave problems raised by misapplication of this capability, but for now we discuss the positive side.

What follows is an exploration of the possible—a look at the devices that could be built once precise molecular control is achieved, and a look at how people might run a manufacturing business based on nanomachines. Try not to think of these sketches as hard-and-fast predictions of precisely how things will be done, but instead as descriptions of capabilities—the sorts of things that can be done once nanotechnology is well in hand. Doubtless there will be better ways to do things than the ways we describe. As usual, references to the 1980s and before are historically accurate; in the rest, the science isn't fiction.

SCENARIO: DESERT ROSE INDUSTRIES

Desert Rose Industries is a diversified wholesale manufacturer of enough furniture, computers, toys, and recreation equipment to have made any twentieth-century captain of industry proud. But if you assembled all Desert Rose employees in front of corporate headquarters, you'd see Carl and Maria Santos standing beside a building the size of a four-bedroom house. This industrial giant is a typical mom-and-pop business, helped along by a network of tele-commuters who handle sales and customer support from homes scattered across North America.

Their friends chide Carl and Maria as "old-fashioned tradition-alists" and tease Maria about abandoning Carl in the factory while she travels to Europe, Asia, South America, and Africa for new business. In the molecular-manufacturing business, familiar personal skills and virtues—honesty, accuracy, good communication—are as important as before. Maria likes to work with the customers. Aided by her S.B. in molecular manufacturing from MIT and her MFA in design, she patiently helps nervous new designers through their first manufacturing experience, and with unflagging courtesy and good humor, handles rush orders, last-second changes, and special orders. Maria's good design ideas and caring personality won them a reputation for being responsive to customer needs. Carl, precise and careful, built their name for accurate manufacturing and delivery on schedule.

Except for Carl's habit of playing Gershwin at full volume with the windows open, the only sounds at the Desert Rose site are the birds along the banks of the stream that winds across the canyon floor; no clanking machinery here. Maria's parents built Desert

Rose Industries out here on an old smelter many miles away from human neighbors. They regraded the land and cleaned up the wastes. Maria adapted a molecular processor to convert heavy-metal contaminants back into stable minerals, and shipped them off to help refill the hole they had originally come from, an old open-pit mine. The desert has mostly healed now, and a few tough trees are spreading along the stream again.

New customers coming up the road for a firsthand look at the manufacturing operations get the full tour: a lunch/meeting room, Maria's office, the manufacturing plant, and the warehouse space for parts and products out back. "The plant" is the largest room, and Carl's pride. Twelve manufacturing ponds and their cooling systems—vats ranging in size from a kitchen sink to a small swimming pool—are where Desert Rose uses nanocomputers and assemblers to do their building work. A plumbers' nightmare of piping runs between the ponds and a triple row of containers with labels like CARBON FEEDSTOCK, PREPARED PLATINUM, SIZE-4 STRUCTURAL FIBERS, and PREFAB MOTORS. Carl keeps a good stock of parts and raw materials on hand, with more in the underground warehouse. Sure, some rare things almost never get used, but having them ready to go is one of Carl's secrets for delivering on time and building precisely to specification. Over on a table are Carl's music system and the computers—descendants of the IBM PCs and Macintoshes of the 1980s—that are used to run the manufacturing process. In a space the size of a large living room, Carl and Maria have all the raw materials and all the production equipment—nanocomputers and assemblers—they need for building almost anything.

Occasionally, Carl and Maria need the services of specialized tools, such as disassemblers, that might exist only in labs. A disassembler works like an archeologist, painstakingly excavating the structure of a molecule, removing atom after atom, in order to record and analyze the molecular structure. Because they work so slowly, noting the position of each molecule, disassemblers aren't used for recycling operations—it would be expensive and pointless to record all this unwanted data. But as tools for analyzing the unknown, they're hard to beat.

Maria found this out when a customer sent her an order for tropically scented furniture and fixtures for his restaurant, but instead of including the software instructions for building the perfume, Maria found a plastic bag full of resinous brown gook with a note saying, "I got this stuff in the tropics. Please make the fabric smell like this." Maria (after sniffing the gook and deciding it smelled surprisingly tropically good) shipped the sample to the lab for chemical analysis by disassembler. The lab sent back software with the molecular description and instructions for building the same scent into the furniture.

Carl usually schedules production very tightly: in every tank, assemblers are building products; every computer is directing work. But this morning, listening to the tone of Maria's voice wafting in from the front office, Carl changes his plans: something important is about to happen. He postpones building orders for video wallpaper and commemorative diamond baseballs, and holds three pools and a computer ready. Minutes later, Maria hurries in, her voice tight and anxious. "Carl, that earthquake down south—they need help. Amanda from the Red Cross is sending the software right now."

To build a product, Desert Rose needs design instructions—computer software—for the assemblers. Carl and Maria have their own software library, but usually they buy or rent what they need, or the customers send their own designs.

The software that Amanda sends contains the specifications to manufacture the emergency equipment: a set of instructions to be run on a standard desktop computer. Within minutes, two copies of the Red Cross software arrive electronically. Before starting the build, Carl meticulously checks to make sure that the master copy and backup copy agree and weren't damaged in transit. If the instructions are complete and correct and properly signed with the Red Cross data stamp, then the desktop computer will communicate these building instructions directly to millions of small computers acting as on-the-job foremen directing the work: nanocomputers.

NANOCOMPUTERS

While the first, primitive assemblers were controlled by changing what molecules are in the solution around the device, getting the speed and accuracy wanted for large-scale manufacturing takes real computation. Carl's setup uses a combination of special-purpose molecule processors and general-purpose assemblers, all controlled and orchestrated by nanocomputers.

Computers back in the 1990s used microelectronics. They worked by moving electrical charge back and forth through conducting paths—wires, in effect—using it to block and unblock the flow of charge in other paths. With nanotechnology, computers are built from molecular electronics. Like the computers of the 1990s, they use electronic signals to weave the patterns of digital logic. Being made of molecular components, though, they are built on a much smaller scale than 1990s computers, and work much faster and more efficiently. On the scale of our simulated molecular world, 1990s computer chips are like landscapes, while nanocomputers are like individual buildings. Carl's desktop PC contains over a trillion nanocomputers, enough to out-compute all the microelectronic computers of the twentieth century put together.

Back in the dark ages of the 1980s, an exploratory engineer proposed that nanocomputers could be mechanical, using sliding rods instead of moving electrons as shown in Figure 8. These molecular mechanical computers were much easier to design than molecular electronic computers would have been. They were a big help in getting some idea of what nanotechnology could do.

Even back then, it was pretty obvious that mechanical computers would be slower than electronic computers. Carl's molecular electronic PC would have been no great surprise, though nobody knew just how to design one. When nanotechnology actually arrived and people started competing to build the best possible computers, molecular electronics won the technology race. Still, mechanical nanocomputers *could* have done all the nanocomputing jobs at Desert Rose: ordinary, everyday molecular manufacturing

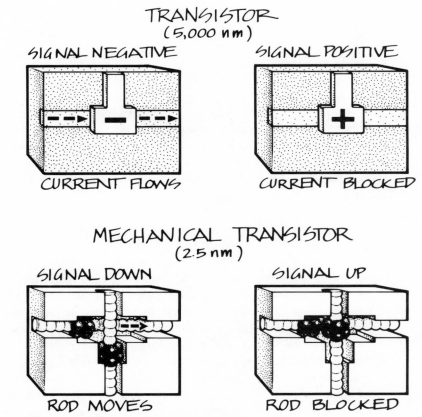

FIGURE 8: MECHANICAL TRANSISTOR

An electronic transistor (above) lets current flow when a negative electric charge is applied and blocks current when a positive charge is applied. The mechanical "transistor" (below) lets the horizontal rod move when the vertical rod is down, and blocks the horizontal rod when the vertical rod is up. Either device can be used to build logic gates and computers.

just doesn't demand the last word in computer performance.

For Carl, the millions of nanocomputers in the milky waters of his building ponds are just extensions of machines on his desk, machines there to help him run his business and deliver products to his customers—or, in the case of the Red Cross emergency, to help provide time-critical emergency supplies. By reserving those three separate ponds, Carl can either build three different kinds of equipment for the Red Cross or use all the ponds to mass-produce the first thing on the Red Cross list: emergency shelters for ten thousand people. The software is ready, the plumbing is fine, the drums of building materials are all topped up, the Special Mix for this job is loaded: the build is ready to start. "Okay," Carl tells the computer, "build Red Cross tents." Computer talks to nanocomputers. In all three pools, nanocomputers talk to assemblers. The build begins.

ASSEMBLING PRODUCTS

Some of the building done at Desert Rose Industries uses assemblers much like the ones we saw in the first hall of the plant tour, back in the simulated molecular world of the Silicon Valley Faire. As seen in simulation, they are big, slow, computer-controlled things moving molecular tools. With the right instructions and machinery to keep them supplied with molecular tools, these general-purpose assemblers can build almost anything. They're slow, though, and take a lot of energy to run. Some of the building uses special-purpose assembly systems in the molecule-processing style, like the systems in the basement we saw in the tour of a simulated molecular factory. The special-purpose systems are all moving belts and rollers, but no arms. This is faster and more efficient, but for quantity orders, cooling requirements limit the speed.

It's faster to use larger, prefabricated building blocks. Desert Rose uses these for most of their work, and especially for rush orders like the one Carl just set up. Their underground warehouse has room-sized bins containing upward of a thousand tons of the most popular building blocks, things like structural fibers. They're made at plants on the West Coast and shipped here by subway for ready use. Other kinds are made on site using the special-purpose assemblers. Carl's main room has several cabinet-sized boxes hooked up to the plumbing, each taking in raw materials, running them through this sort of specialized molecular machinery, and pumping out a milky syrup of product. One syrup contains motors, another one contains computers, and another is full of microscopic plug-in light sources. All go into tanks for later use.

Now they're being used. The mix for the Red Cross tent job is mostly structural fiber stronger than the old bulletproof-vest materials. Other building blocks also go in, including motors, computers, and dozens of little struts, angle brackets, and doohickies. The mix would look like someone had stirred together the parts from a dozen toy sets, if the parts were big enough to see. In fact, though, the largest parts would be no more than blurry dots, if you saw one under a normal optical microscope.

The mix also contains block-assemblers, floating free like everything else. These machines are big, about like an office building in our simulation view with the standard settings. Each has several jointed arms, a computer, and several plugs and sockets. These do the actual construction work.

To begin the build, pumps pour the mix into a manufacturing pond. The constant tumbling motions of microscopic things in liquids would be too disorganized for building anything so large as a tent, so the block-assemblers start grabbing their neighbors. Within moments, they have linked up to form a framework spread through the liquid. Now that they are plugged together, they divide up jobs, and get to work. Instructions pour in from Carl's desktop computer.

The block-assemblers use sticky grippers to pull specific kinds of building blocks out of the liquid. They use their arms to plug them together. For a permanent job, they would be using blocks that bond together chemically and permanently. For these temporary tents, though, the Red Cross design uses a set of standard blocks that are put together with amazingly ordinary fasteners: these blocks have snaps, plugs, and screws, though of course the parts are atomically perfect and the threads on the screws are single helical rows of atoms. The resulting joints weaken the tent's structure somewhat, but who cares? The basic materials are almost a hundred times stronger than steel, so there is strength to waste if it makes manufacturing more convenient.

Fiber segments snap together to make fabrics. Some segments contain motors and computers, linked by fibers that contain power and data cables. Struts snap together with more motors and computers to make the tent's main structures. Special surfaces are made of special building blocks. From the human perspective, each tent is a lightweight structure that contains most of the conveniences and comforts of an apartment: cooking facilities, a bathroom, beds, windows, air-conditioning, specially modified to meet the environmental demands of the quake-stricken country. From a builder's perspective, especially from a nanomachine's point of view, the tent is just structure slapped together from a few hundred kinds of prefab parts.

In a matter of seconds, each block-assembler has put together a few thousand parts, and its section of the tent is done. In fact, the whole thing is done: many trillions of hands make light work. A crane swings out over the pond and starts plucking out tent packages as fresh mix flows in.

Maria's concern has drawn her back to the plant to see how the build is going. "It's coming along," Carl reassures her. "Look, the first batch of tents is out." In the warehouse, the first pallet is already stacked with five layers of dove-gray "suitcases": tents dried and packed for transport. Carl grabs a tent by the handle and lugs it out the door. He pushes a tab on the corner labeled "Open," and it takes over a minute to unfold to a structure a half-dozen paces on a side. The tent is big, and light enough to blow away if it didn't cling to the ground so tightly. Maria and Carl tour the tent, testing the appliances, checking the construction of furniture: everything is extremely lightweight compared to the bulk-manufactured goods of the 1990s, tough but almost hollow.

Like the other structures, the walls and floors are full of tiny motors and struts controlled by simple computers like the ones used in twentieth-century cars, televisions, and pinball machines. They can unfold and refold. They can also flex to produce sound like a high-quality speaker, or to absorb sound to silence outdoors racket. The whole three-room setup is small and efficient, looking like a cross between a boat cabin and a Japanese business hotel room. Outside, though, it is little more than a box. Maria shakes her head, knowing full well what architects can do these days when they try to make a building really fit its site. *Oh well*, she thinks, *These won't be used for long.*

"Well, that looks pretty good to me," says Carl with satisfaction. "And I think we'll be finished in another hour."

Maria is relieved. "I'm glad you had those pools freed up so fast."

By three o'clock, they've shipped three thousand emergency shelters, sending them by subway. Within half an hour, tents are being set up at the disaster site.

BEHIND THE SCENES AND AFTERWARD

Desert Rose Industries and other manufacturers can make almost anything quickly and at low cost. That includes the tunneling machines and other equipment that made the subway system they use for shipping. Digging a tunnel from coast to coast now costs less than digging a single block under New York City used to. It wasn't expensive to get a deep-transit terminal installed in their basement. Just as the tents aren't mere bundles of canvas, these subways aren't slow things full of screeching, jolting metal boxes. They're magnetically levitated to reach aircraft speeds—as experimental Japanese trains were in the late 1980s—making it easy for Carl and Maria to give their customers quick service. There's still a road leading to the plant, but nobody's driven a truck over it for years.

They only take in materials that they will eventually ship out in products, so there's nothing left over, and no wastes to dump. One corner of the plant is full of recycling equipment. There are always some obsolete parts to get rid of, or things that have been damaged and need to be reworked. These get broken down into simpler molecules and put back together again to make new parts.

The gunk in the manufacturing ponds is water mixed with particles much finer than silt. The particles—fasteners, computers, and the rest—stay in suspension because they are wrapped in molecular jackets that keep them there. This uses the same principle as detergent molecules, which coat particles of oily dirt to float them away.

Though it wouldn't be nutritious or appetizing, you could drink the tent mix and be no worse for it. To your body, the parts and their jackets, and even the nanomachines, would be like so many bits of grit and sawdust. (Grandma would have called it roughage.)

Carl and Maria get their power from solar cells in the road, which is the only reason they bothered having it paved. In back of their plant stands what looks like a fat smokestack. All it produces, though, is an updraft of clean, warm air. The darkly paved road, baking in the New Mexico sun, is cooler than you might expect: it soaks up solar energy and makes electricity, instead of just heat. Once the power is used, it turns back into heat, which has to go somewhere. So the heat rises from their cooling tower instead of the road, and the energy does useful work on the way.

Some products, like rocket engines, are made more slowly and in a single piece. This makes them stronger and more permanent. The tents, though, don't need to be superstrong and are just for temporary use. A few days after the tents go up, the earthquake victims start to move out into new housing (permanent, better-looking, and *very* earthquake-resistant). The tents get folded and shipped off for recycling.

Recycling things built this way is simple and efficient: nanomachines just unscrew and unsnap the connectors and sort the parts into bins again. The shipments Desert Rose gets are mostly recycled to begin with. There's no special labeling for recycled materials, because the molecular parts are the same either way.

For convenience (and to keep the plant small), Carl and Maria get most of their parts prefabricated, even though they can make almost anything. They can even make more production equipment. In one of their manufacturing ponds, they can put together a new cabinet full of special-purpose assemblers. They do this when they want to make a new type of part in-house. Like parts, the part-assemblers are made by special-purpose assemblers. Carl can even make big vats in medium-size vats, unfolding them like tents.

If Desert Rose Industries needed to double capacity, Carl and Maria could do it in just a few days. They did this once for a special order of stadium sections. Maria got Carl to recycle the new building before its shadow hurt their cactus garden.

FACTORY FACTORIES

In the Desert Rose Industries scenario, manufacturing has become cheap, fast, clean, and efficient. Using fast, precise machines to handle matter in molecular pieces makes it easy for nanotechnology to be fast, clean, and efficient. But for it to be *cheap*, the manufacturing equipment has to be cheap.

The Desert Rose scenario shows how this can work. Molecular-manufacturing equipment can be used to make all the parts needed to build more molecular manufacturing equipment. It can even build the machines needed to put the parts together. This resembles an idea developed by NASA for a self-expanding manufacturing complex on the Moon, but made faster and simpler using molecular machines and parts.

REPLICATORS

In the early days of nanotechnology, there won't be as many different kinds of machines as there are at Desert Rose. One way to build a lot of molecular-manufacturing equipment in a reasonable time would be to make a machine that can be used to make a copy of itself, starting with special but simple chemicals. A machine able to do this is called a "replicator." With a replicator and a pot full of the right fuel and raw materials, you could start with one machine, then have two, four, eight, and so on.

This doubling process soon makes enough machines to be useful. The replicators—each including a computer to control it and a general-purpose assembler to build things—could then be used to make something else, like the tons of specialized machines needed to set up a Desert Rose manufacturing plant. At that point, the replicators could be discarded in favor of those more efficient machines.

Replicators are worth a closer look, though, because they show how quickly molecular manufacturing systems can be used to build more manufacturing equipment. Figure 9 shows a design described in Stanford University course CS 404 in the spring of 1988. If we were in one of our standard simulation views, the submicroscopic device at the top of the picture would be like a huge tank, three stories tall when lying on its side. Most of its interior is taken up by a tape memory system that tells how to move the arm to build all the parts of the replicator, except the tape itself. The tape gets made by

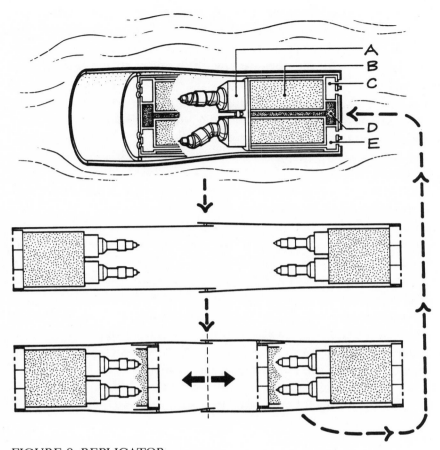

FIGURE 9: REPLICATOR

A replicator would be able to build copies of itself when supplied with fuel and raw materials. In the diagram, (A) contains a nanocomputer, (B) a library of stored instructions, (C) contains machinery that takes in fuel and produces electric power, (D) is a motor, and (E) contains machinery that prepares raw materials for use. (All volumes follow calculations presented in a class at Stanford.) The lower diagrams illustrate steps in a replication cycle, showing how the working space is kept isolated from the external liquid, which provides the needed fuel and raw-material molecules. Replicators of this sort are useful as thought experiments to show how nanomachines can produce more nanomachines, but specialized manufacturing equipment would be more efficient in practice.

a special tape-copying machine. At the right-hand end of the replicator are pores for bringing in fuel and raw-material molecules, and
machinery for processing them. In the middle are computer-controlled arms, like the ones we saw on the plant trip. These do most
of the actual construction.

The steps in the cycle—using a copy to block the tube, beginning
a fresh copy, then releasing the old one—illustrate one way for a
machine to build a copy of itself while floating in a liquid, yet doing
all its construction work inside, in vacuum. (It's easier to design for
vacuum, and this is exploratory-engineering work, so easier design is
better design.) Calculations suggest that the whole construction cycle
can be completed in less than a quarter hour, since the replicator
contains about a billion atoms, and each arm can handle about a
million atoms per second. At that rate, one device can double and
double again to make trillions in about ten hours.

Each replicator just sits in a chemical bath, soaking up what it
needs and making more replicators. Eventually, either the special
chemicals run out, or other chemicals are added to signal them to
do something else. At that point, they can be reprogrammed to produce anything else you please, so long as it can be extruded from the
front. The products can be long, and can unfold or be pieced together to make larger objects, so the size of these initial replicators—
smaller than a bacterium—would be only a temporary limitation.

General Assemblers

From the molecular manipulators and primitive assemblers described in the last chapter, the most likely path to nanotechnology
leads to assemblers with more and more general capabilities. Still,
efficiency favors special-purpose machines, and the Desert Rose scenario didn't make much use of general assemblers. Why bother making general-purpose assemblers in the first place?

To see the answer, turn the question around and ask, Why *not*
build such a tool? There is nothing outstandingly difficult about a
general assembler, as molecular machinery goes. It will just be a device with good, flexible positional control and a system to feed it a

variety of molecular tools. This is a useful, basic capability. General-purpose assemblers could always be replaced by a lot of specialized devices, but to build those specialized devices in the first place, it makes sense to come up with a more flexible, general-purpose system that can just be reprogrammed.

So, general purpose machines are likely to find use in making short production runs of more specialized devices. Ralph Merkle, a computers and security expert at Xerox Palo Alto Research Center, sees this as paralleling the way manufacturing works today: "General purpose devices could do many tasks, but they'll do them inefficiently. For any given task, there will be one or a few best ways of doing it, and one or a few special-purpose devices that are finely tuned to do that one task. Nails aren't made by a general-purpose machine shop, they're made by nail-making machines. Making nails with a general-purpose machine shop would be more expensive, more difficult, and more time-consuming. Likewise, in the future we won't see a proliferation of general-purpose self-replicating systems, we'll see specialization for almost every task."

WHAT WILL THESE CAPABILITIES MAKE POSSIBLE?

We've surveyed a lot of devices: assemblers of various flavors, nano-computers, disassemblers, replicators, and others. What's important about these is not the exact distinctions between them, but the capabilities that they will give and the effects they will have on human lives. Again, we are suspending discussion of potential misapplications until later.

If we tease apart the implications of what we've seen in the Desert Rose scenario, we can analyze some of the key impacts of molecular manufacturing in industry, science, and medicine.

TECHNOLOGY AND INDUSTRY

At its base, nanotechnology is about molecular manufacturing, and manufacturing is the basis of much of today's industry. This is why Desert Rose made a good starting point for describing the possibilities of a nanotechnological world. From an industrial perspective, it makes sense to think of nanotechnology in terms of products and production.

New Products: Today, we handle matter crudely, but nanotechnology will bring thorough control of the structure of matter, the ability to build objects to atom-by-atom specifications. This means being able to make almost anything. By comparison, even today's range of products will feel very limited. Nanotechnology will make possible a huge range of new products, a range we can't envision today. Still, to get a feel for what is possible, we can look at some easily imagined applications.

Reliable Products: Today, products often fail, but for failures to occur—for a wing to fall off an airplane, or a bearing to wear out— a *lot* of atoms have to be out of place. In the future, we can do better. There are two basic reasons for this: better materials and better quality control, both achieved by molecular manufacturing. By using materials tens of times stronger than steel, as Desert Rose did, it will be easy to make things that are very strong, with a huge safety margin. By building things with atom-by-atom control, flaws can be made very rare and extremely small—nonexistent, by present standards.

With nanotechnology, we can design in big safety margins and then manufacture the design with near-perfection. The result will be products that are tough and reliable. (There will still be room for bad designs, and for people who wish to take risks in machines that balance on the edge of disaster.)

Intelligent Products: Today, we make most things from big chunks of metal, wood, plastic, and the like, or from tangles of fibers. Objects made with molecular manufacturing can contain trillions of microscopic motors and computers, forming parts that work together to do something useful. A climber's rope can be made of fibers that slide around and reweave to eliminate frayed spots. Tents can be made of parts that slide and lock to turn a package into a building. Walls and furniture can be made to repair themselves, instead of passively deteriorating.

On a mundane level, this sort of flexibility will increase reliability and durability. Beyond this, it will make possible new products with abilities we never imagined we needed so badly. And beyond even this, it will open new possibilities for art.

Inexpensive Production: Today, production requires a lot of labor, either for making things or for building and maintaining machines that make things. Labor is expensive, and expensive machines make automation expensive, too. In the Desert Rose scenario, we got a glimpse of how molecular manufacturing can make production far less expensive than it is today. This is perhaps the most surprising conclusion about nanotechnology, so we'll take a closer look at it in the next chapter.

Clean Production: Today, our manufacturing processes handle matter sloppily, producing pollution. One step puts stuff where it shouldn't be; the next washes it off the product and into the water supply. Our transportation system worsens the problem as unreliable trucks and tankers spill noxious chemicals over the land and sea. Everything is expensive, so companies skimp on even the half-effective pollution controls that we know how to build.

Nanotechnology will mean greater control of matter, making it easy to avoid pollution. This means that a little public pressure will go a long way toward a cleaner environment. Likewise, it will make it easy to increase efficiency and reduce resource requirements. Products, like the Red Cross tents at Desert Rose, can be made of snap-together, easily recyclable parts. Sophisticated products could even be made from biodegradable materials. Nanotechnology will make it easy to attack the causes of pollution at their technological root.

Nanotechnology will have great applications in the field of industry, much as transistors had great applications in the field of vacuum-tube electronics, and democracy had great applications in the field of monarchy. It will not so much *advance* twentieth-century industry as *replace* it—not all at once, but during a thin slice of historical time.

SCIENCE

Chemistry: Today, chemists work with huge numbers of molecules and study them using clever, indirect techniques. Making a new molecule can be a major project, and studying it can be another. Molecular manufacturing will help chemists make what they want to study, and it will help them make the tools they need to study it. Nanoinstruments will be used to prod, measure, and modify molecules in a host of ways, studying their structures, behaviors, and interactions.

Materials: Today, materials scientists make new superconductors, semiconductors, and structural materials by mixing and crushing and baking and freezing, and so forth. They dream of far more structures than they can make, and they stumble across more things than they plan. With molecular manufacturing, materials science can be much more systematic and thorough. New ideas can be tested because new materials can be built according to plan (rather than playing around, groping for a recipe).

This need not rule out unexpected discoveries, since experiments—even blind searches—will go much faster. A few tons of raw materials would be enough to make a billion billion samples, each a cubic micron in size. In all of history so far, materials scientists have never tested so many materials. With nanoinstruments and nanocomputers, they could. One laboratory could then do more than all of today's materials scientists put together.

Biology: Today, biologists use a host of molecular devices borrowed from biology to study biology. Many of these can be viewed as molecular machines. Nanotechnology will greatly advance biology by providing better molecular devices, better nanoinstruments. Some cells have already been mapped in amazing molecular detail, but biology still has far to go. With nanoinstruments (including molecule-by-molecule disassemblers), biologists will at last be able to map cells completely and study their interactions in detail. It will become easy not only to find molecules in cells, but to learn what they do. This will help in understanding disease and the molecular requirements for health, enormously advancing medicine.

Computation: Today, computers range from a million to a billion times faster than an old desktop adding machine, and the results have been revolutionary for science. Every year, more questions can be answered by calculations based on known principles of physics. The advent of nanocomputers—even slow, miserable, mechanical nanocomputers—will give us practical machines with a *trillion* times the power of today's computers (essentially by letting us package a trillion computers in a small space, without gobbling too much money or energy.) The consequences will again be revolutionary.

Physics: The known principles of physics are adequate for understanding molecules, materials, and cells, but not for understanding phenomena on a scale that would still be submicroscopic if atoms were the size of marbles. Nanotechnology can't help here directly, but it can provide manufacturing facilities that will make huge particle accelerators economical, where today they strain national budgets.

More generally, nanotechnology will help science wherever precision and fine details are important. Science frequently proceeds by trying small variations in almost identical experiments, comparing the results. This will be easier when molecular manufacturing can make two objects that are identical, molecule by molecule. In some areas, today's techniques are not only crude, but destructive. Archaeological sites are unique records of the human past, but today's techniques throw away most information during the dig, by accident. Future archaeologists, able to sift soil not speck by speck but molecule by molecule, will be grateful indeed to those archaeologists who today leave some ground undisturbed.

MEDICINE

Of all the areas where the ability to manufacture new tools is important, medicine is perhaps the greatest. The human body is intricate, and that intricacy extends beyond the range of human vision, beyond microscopic imaging, down to the molecular scale. "Molecular medicine" is an increasingly popular term today, but medicine today has only the simplest of molecular tools. As biology uses nanoinstruments to learn about disease and health, we will learn the physical requirements for restoring and maintaining health. And with this knowledge will come the tools needed to satisfy those requirements—tools ranging from improved pharmaceuticals to devices able to repair cells and tissues through molecular surgery.

Advanced medicine will be among the most complex and difficult applications of nanotechnology. It will require great knowledge, but nanoinstruments will help gather this knowledge. It will pose great engineering challenges, but computers of trillionfold greater power will help meet those challenges. It will solve medical problems on which we spend billions of dollars today, in hopes of modest improvements.

Today, modern medicine often means an expensive way to prolong misery. Will nanomedicine be more of the same? Any reader over the age of, say, thirty knows how things start to go wrong: an ache here, a wrinkle there, the loss of an ability. Over the decades, the physical quality of life declines faster and faster—the limits of what the body can do become stricter—until the limits are those of a hospital bed. The healing abilities we have when young seem to fade away. Modern medical practice expends the bulk of its effort on such things as intensive-care units, dragging out the last few years of life without restoring health.

Truly advanced medicine will be able to restore and supplement the youthful ability to heal. Its cost will depend on the cost of producing things more intricate than any we have seen before, the cost of producing computers, sensors, and the like by the trillions. To understand the prospects for medicine, like those for science and industry, we need to take a closer look at the cost of molecular manufacturing.

SEVEN

The Spiral
of Capability

In earlier chapters, we have stepped forward and backward through time. The last step was a big one, leaping from small laboratory devices to the high-capacity industrial facility of the Desert Rose scenario. Our narrative crossed this gap in a single leap, but the world won't. To understand how nanotechnology might unfold, it makes sense to look at some of its easier and more difficult applications. The result won't be a timetable, or even a series of milestones, but it should give a better picture of what we can expect as nanotechnology develops from simple, crude, costly beginnings to a state of greater sophistication and lower cost.

IMPROVING QUALITY

Molecular manufacturing will make better products possible. We're likely to see some early applications in at least two areas: stronger materials and faster computers. Strong materials are simple, and will be hard to pass up. Computers are more complex, but the payoff will be enormous.

COMPUTERS

The computer industry has been under steady pressure to make computer chips ever smaller. As sizes have shrunk, costs have fallen, while efficiency and capabilities have increased. The pressure to continue this process pushes in the direction of nanotechnology; it may even be one of the major motivations behind developing the technology.

John Walker, a founder of Autodesk, explains: "Even technologies with enormous potential can lie dormant unless there are significant payoffs along the way to reward those who pioneer them. That's one of the reasons integrated circuits developed so rapidly; each advance found an immediate market willing to apply it and enrich the innovator that created it.

"Does molecular engineering have this kind of payoff? I think it does. Remembering that we may be less than ten years away from 'hitting the wall' as far as scaling our existing electronics goes, a great deal of research is presently going on in the area of molecular and quantum electronics. The payoff is easy to calculate: You can build devices one thousand times faster, more energy-efficient, and cheaper than those we're currently using—at least one hundred times better than exotic materials being considered to replace silicon when it reaches its limits."

Federico Capasso, head of the Quantum Phenomena and Device Research Department at AT&T Bell Labs, agrees that electronics researchers will keep pushing for smaller devices once silicon's potential has been reached. He explains that "at some point we will reach difficulties: some people say at a hundred-fifty nanometers, others think it's beyond that. What will happen then? It's hard to think that the electronics industry will say, 'Stop here. We'll stop evolving because we can't shrink the device.' From an economic point of view, in order to survive, an industry has to innovate continuously."

The computer industry's push toward devices of molecular size has an air of inevitability. Today's researchers struggle to build molecular electronics using bulk techniques, with no products yet in sight; with molecular manipulators, they will finally have the tools they need for fast and accurate experimentation. Once successful designs are developed, packaged, and tested, the pressure will be on to learn to make them in quantity at low cost. The competitive pressures will be fierce, because advanced molecular electronics will be *orders of magnitude* better than today's integrated circuits, ultimately enabling the construction of computers with trillionfold greater capability.

STRONG, LIGHTWEIGHT STRUCTURES

At the opposite extreme from molecular electronics—complex and at first worth billions of dollars per gram—are structural materials: worth only dollars per kilogram in most applications, but much simpler in structure. Once molecular manufacturing becomes inexpensive, structural materials will be important products.

These materials play a central role in almost everything around us, from cars and aircraft to furniture and houses. All of these objects get their size, shape, and strength from a structural skeleton of some sort. This makes structural materials a natural place to begin in understanding how nanotechnology can improve products.

Cars today are mostly made of steel, aircraft of aluminum, and buildings and furniture largely of steel and wood. These materials have a certain "strength-to-weight ratio" (more properly, a strength-to-density ratio). To make cars stronger, they'd have to be heavier; to make them lighter, they'd have to be weaker. Clever design can change this relationship a little, but to change it a lot requires a change of materials.

Making something heavy is easy: just leave a hollow space, then fill it with water, sand, or lead shot. Making something light and strong is harder, but often important. Automakers try to make cars lightweight, aircraft manufacturers try harder, and with spacecraft manufacturers it is an obsession. Reducing mass saves materials and energy.

The strongest materials in use today are mostly made of carbon. Kevlar, used in racing sails and bulletproof vests, is made of carbon-rich molecular fibers. Expensive graphite composites, used in tennis rackets and jet aircraft, are made using pure-carbon fibers. Perfect fibers of carbon—both graphite and diamond—would be even better, but can't be made with today's technology. Once molecular manufacturing gets rolling, though, such materials will be commonplace and inexpensive.

What will these materials be like? To picture them, a good place to start is wood. The structure of wood can vary from extremely light and porous, like balsa wood, to denser structures like oak. Wood is made by molecular machinery in plants from carbon-rich polymers, mostly cellulose. Molecular manufacturing will be able to make materials like these, but with a strength-to-weight ratio about a hundred times that of mediocre steel, and tens of times better than the best steel. Instead of being made of cellulose, these materials will be made of carbon in forms like diamond.

Diamond is emphasized here not because it is shiny and expensive, but because it is strong and potentially cheap. Diamond is just carbon with properly arranged atoms. Companies are already learning to make it from natural gas at low pressure. Molecular manufacturing will be able to make complex objects of the stuff, built lighter than balsa wood but stronger than steel.

Products made of such materials could be startling by our present standards. Objects could be made that are identical in size and shape to those we make today, but simultaneously stronger and 90 percent lighter. This is something to keep in mind next time you're lugging a heavy object around. (If something needs weight to hold it in place, it would be more convenient to add this ballast when the thing is in its proper location than to build in the extra weight permanently.)

Better structural materials will make aircraft lighter, stronger, and more efficient, but will have the greatest effect on spacecraft. Today, spacecraft can barely reach orbit with both a safety margin and a cargo. To get there at all, they have to drop off parts like boosters and tanks along the way, shedding weight. With strong materials, this will change: as in the space-travel-for-business scenario in Chapter 1, spacecraft will become more like aircraft are today. They will be rugged and reliable, and strong enough and light enough to reach space in one piece.

QUICKENING DEVELOPMENT

In some areas of high technology—spaceflight has been a notorious example—it takes years, even decades, to try a new idea. This makes progress slow to a crawl. In other areas—software has been a shining example—new ideas can be tested in minutes or hours. Since the Space Shuttle design was frozen, personal computer software has come into existence and gone through several generations of commercial development, each with many cycles of building and testing.

FAST, INEXPENSIVE TESTING

Even in the days of the first operational molecular manipulators, experimentation is likely to be reasonably fast. Individual chemical steps can take seconds or less. Complex molecular objects could be built in a matter of hours. This will let new ideas be put into practice almost as fast as they can be designed.

Later assemblers will be even faster. At a millionth of a second per step, they will approach the speed of computers. And, as nanotechnology matures, experimenters will have more and more molecular instruments available to help them find out whether their devices work or not. Fast construction and fast testing will encourage fast progress.

At this point, the cost of materials and equipment for experiments will be trivial. No one today can afford to build Moon rockets on a hobby budget, but they can afford to build software, and many useful programs have been the result. There is no *economic* reason why nanomachines couldn't eventually be built with a hobby-size budget, though there are reasons—to be discussed in later chapters—for wanting to place limits on what can be built.

EARLY SIMPLICITY

Finally, established technologies are always pushing up against some limit; the easy opportunities have generally been exploited. In many fields, the limits are those of the properties of the materials used and the cost and precision of manufacturing. This is true for computers, for spacecraft, for cars, blenders, and shoes. For software, the limits are those of computer capacity and of sheer complexity (which is to say, of human intelligence). After molecular manufacturing develops certain basic abilities, a whole set of limits will fall, and a whole range of developments will become possible. Limits set by materials properties, and by the cost and precision of manufacturing, will be pushed way back. Competition, easy opportunities, and fast, low-cost experimentation should combine to yield an explosion of new products.

This does not mean *immediately*, and it does not apply to *all* imaginable nanotechnologies. Some technologies are imaginable and clearly feasible, yet dauntingly complex. Still, the above considerations suggest that a wide range of advances could happen at a brisk pace. The main bottleneck might seem to be a shortage of knowledgeable designers—hardly anyone knows both chemistry and mechanical design—but improving computer simulations will help. These simulations will let engineers tinker with molecular-machinery designs, absorbing knowledge of chemical rules without learning chemistry in the usual sense.

CLIMBING COMPLEXITY

Making familiar products from improved materials will increase their safety, performance, and usefulness. It will also present the simplest

	Space	Computers	Nanotechnology
Precursor science and technologies	Physics Sounding rockets	Mathematics Electronics	Theoretical chemistry Chemical synthesis
Crucial advance	Teams combine and improve technologies	Teams combine and improve technologies	Teams combine and improve technologies
Threshold capability	First satellite	First computer	First assembler
Early practical applications	Weather, spy, and communication satellites	Scientific calculations Payroll calculations	Molecular sensors Molecular computing
Breakthrough capability	Routine, inexpensive spaceflight	Powerful mass-market desktop computers	Powerful, inexpensive molecular manufacturing
Further projected developments	Lunar base, Mars exploration	Widespread electronic publishing	New medical abilities New, inexpensive products
More advanced developments	Mining, development, settlement of solar system	Major automation of engineering design	Help with computer goals Environmental cleanup
Yet more advanced developments	Interstellar flight and settlement feasible	Trillionfold computer power	Help with space goals General tissue repair

engineering task. A greater change, though, will result from unfamiliar products made possible by new manufacturing methods. In talking about unfamiliar products, a hard-to-answer question arises: What will people want?

Products are typically made because their recipients want them. In our discussions here, if we describe something that people won't want, then it probably won't get built, and if it does get built, it will soon disappear. (The exceptions—fraud, coercion, persistent mistakes—are important, but in other contexts.) To anchor our discussion, it makes sense to look not at totally new products, but instead at new features for old products, or new ways to provide old services. This approach won't cover more than a fraction of what is possible, but will start from something sensible and provide a springboard for the imagination.

As usual, we are describing possibilities, not making predictions. The possibilities focused on here arise from more complex applications of molecular manufacturing—nanotechnological products that contain nanomachines when they are finished. Earlier, we discussed strong materials. Now, we discuss some smart materials.

Smart Materials

The goal of making materials and objects smart isn't new: researchers are already struggling to build structures that can sense internal and environmental conditions and adapt themselves appropriately. There is even a *Journal of Intelligent Material Systems and Structures.* By using materials that can adapt their shapes, sometimes hooked up to sensors and computers, engineers are starting to make objects they call "smart." These are the early ancestors of the smart materials that molecular manufacturing will make possible.

Today, we are used to having machines with a few visible moving parts. In cars, the wheels go around, the windshield wipers go back and forth, the antenna may go up and down, the seat belts, mirrors, and steering wheel may be motor-driven. Electric motors are fairly small, fairly inexpensive, and fairly reliable, so they are fairly common. The result is machines that are fairly smart and flexible, in a clumsy, expensive way.

In the Desert Rose scenario, we saw "tents" being assembled from trillions of submicroscopically small parts, including motors, computers, fibers, and struts. To the naked eye, materials made from these parts could seem as smooth and uniform as a piece of plastic, or as richly textured as wood or cloth—it is all a matter of the arrangement and appearance of the submicroscopic parts. These motors and other parts cost less than a trillionth of a dollar apiece. They can be quite reliable, and good design can make systems work smoothly even if 10 percent of a trillion motors burn out. Likewise for motor-controlling computers and the rest. The resulting machines can be very smart and flexible, compared to those of today, and inexpensive, too.

When materials can be full of motors and controllers, whole chunks of material can be made flexible and controllable. The applications should be broad.

SCENARIO: SMART PAINT

Surfaces surround us, and human-made surfaces—walls, roofs, and pavement—cover huge areas that matter to people. How can smart materials make a difference here?

> The revolution in technology has come and gone, and you want to repaint your walls. Breathing toxic solvents and polluting water by washing brushes have passed into history, because paint has been replaced with smarter stuff. The mid-twentieth century had seen considerable progress in paints, especially the development of liquids that weren't quite liquid—they would spread with a brush, but didn't (stupidly) run and drip under their own weight. This was an improvement, but the new material, "paperpaint," is even more cooperative.
>
> Paperpaint comes in a box with a special trowel and pen. The paperpaint itself is a dry block that feels a lot like a block of wood.

Following the instructions, you use the pen to draw a line around
the edge of the area you want to paint, putting an X in the middle
to show where you want the paint to go on; the line is made of
nontoxic disappearing ink, so you can slop it around without stain-
ing anything. Using the trowel, you slice off a hunk of paperpaint—which is easy, because it parts like soft butter to the trowel,
even though it behaves like a solid to everything else. Very high IQ
stuff, that.

Next, you press the hunk against the X and start smoothing it
out with the trowel. Each stroke spreads a wide swath of paperpaint, much wider than the trowel, but always staying within the
inked line. A few swipes spreads it precisely to the edges, where-
upon it smooths out into a uniform layer. Why doesn't it just
spread itself? Experience showed that customers didn't mind the effort of making a few swipes and preferred the added control.

The paperpaint consists of a huge number of nanomachines
with little wheels for rolling over one another and little sticky pads
for clinging to surfaces. Each has a simple, stupid computer on
board. Each can signal its neighbors. The whole mass of them
clings together like an ordinary solid, but they can slip and slide in
a controlled way when signaled. When you smooth the trowel over
them, this contact tells them to get moving and spread out. When
they hit the line, this tells them to stop. If they don't hit a line,
they go a few handbreadths, then stop anyway until you trowel
them again. When they encounter a line on all sides, word gets
around, and they jostle around to form a smooth, uniform layer.
Any that get scraped off are just so much loose dust, but they stick
together quite well.

This paint-stuff doesn't get anything wet, doesn't stain, and
clings to surfaces just tightly enough to keep it from peeling off accidentally. If some experimentally minded child starts digging with
a stick, makes a tear, and peels some off, it can be smoothed back
again and will rejoin as good as new. The child may eat a piece,
but careful regulation and testing has ensured that this is no worse
than eating plain paper, and safer than eating a colorful Sunday
newspaper page.

Many refinements are possible. Swipes and pats of the trowel
could make areas thicken or thin, or bridge small holes (no more
Spackling!). With sufficiently smart paperpaint, and some way to

indicate what it should do, you can have your choice of textures. Any good design will be washable, and a better design would shed dirt automatically using microscopic brushes.

Removal, of course, is easy: either you rip and peel (no scraping needed), or find that trowel, set the dial on the handle to "Strip," and poke the surface a few times. Either way, you end up with a lump ready to pitch into the recycling bin and the same old wall you started with, bared to sight again.

POWER PAINT

Perhaps no product will ever be made exactly like the smart paint just described. It would be disappointing if something better couldn't be made by the time smart paint is technologically possible. Still, paperpaint gives a feel for some of the features to expect in the new smart products, features such as increased flexibility and better control. Without loading yet more capability into our paint (though there is no reason why one *couldn't*), let's take a look at some other smart properties one might want in a surface.

External walls, roofs, and paving surfaces are exposed to sunlight, and sunlight carries energy. A proven ability of molecular machinery is the conversion of sunlight to stored energy: plants do it every day. Even now, we can make solar cells that convert sunlight into electricity at efficiencies of 30 percent or so. Molecular manufacturing could not only make solar cells much cheaper, but could also make them tiny enough to be incorporated into the mobile building blocks of a smart paint.

To be efficient, this paint would have to be dark—that is, would have to absorb a lot of light. Black would be best, but even light colors could generate some power, and efficiency isn't everything. Once the paint was applied, its building blocks would plug together to pool their electrical power and deliver it through some standard plug. A thicker, tougher form of this sort of material could be used to resurface pavement, generate power, and transmit it over large distances. Since smart solar-cell pavement could be designed for improved traction and a similar roofing material could be designed for amazing leak-resistance, the stuff should be popular.

On a sunny day, an area just a few paces on a side would generate a kilowatt of electrical power. With good batteries (and enough repaved roads and solar-cell roofing), present demands for electrical power could be met with *no* coal burning, *no* oil imports, *no* nuclear power, *no* hydroelectric dams, and *no* land taken over for solar-power generation plants.

PRETTY PAINT, ACOUSTIC PAINT

The glow of fireflies and deep-sea fish shows that molecular devices can convert stored chemical energy into light. All sorts of common devices show that electricity can be converted to light. With molecular manufacturing, this conversion can be done in thin films, with control over the brightness and color of each microscopic spot. This could be used for diffuse lighting—ceiling paperpaint that glows. With more elaborate control, this would yield the marvel (horror?) of video wallpaper.

With today's technology, we are used to displays that glow. With molecular manufacturing, it will be equally easy to make displays that just change color, like a printed page with mobile ink. Chameleons and flatfish change color by moving colored particles around, and nanomachines could do likewise. On a more molecular level, they could use tunable dyes. Live lobsters are a dark grayish green, but when cooked turn bright red. Much of this change results from the "retuning" of a dye molecule that is bound in a protein in the live lobster but released by heat. This basically mechanical change alters its color; the same principle can be used in nanomachines, but reversibly.

How a surface appears depends on how it reflects or emits light. Nanomachines and nanoelectronics will be able to control this within wide limits. They will be able to do likewise for sound, by controlling how a surface moves. In a stereo system, a speaker is a movable surface, and nanomachines are great for making things move as desired. Making a surface emit high-quality sound will be easy. Almost as easy will be surfaces that actively flex to absorb sound, so that the barking dog across the street seems to fade away.

SMART CLOTH

Looking further at the human environment, we find a lot of cloth and related materials, such as carpeting and shoes. The textile industry was at the cutting edge of the first industrial revolution, and the next industrial revolution will have its effects on textiles.

With nanotechnology, even the finest textile fibers could have sensors, computers, and motors in their core at little extra cost. Fabrics could include sensors able to detect light, heat, pressure, moisture, stress, and wear, networks of simple computers to integrate this data, and motors and other nanomechanisms to respond to it. Ordinary, everyday things like fabric and padding could be made responsive to a person's needs—changing shape, color, texture, fit, and so forth—with the weather and a person's posture or situation. This process could be slow, or it could be fast enough to respond to a gesture. One result would be genuine one-size-fits-all clothing (give or take child sizes), perfectly tailored off the rack, warm in winter, cool and dry in summer; in short, nanotechnology could provide what advertisers have only promised. Even bogus advertising gives a clue to human desires.

Throughout history, the human race has pursued the quest for comfortable shoes. With fully adjustable materials, the seemingly impossible goal of having shoes that both look good and feel good should finally be achieved. Shoes could keep your feet dry, and warm except in the Arctic, cool except in the tropics, and as comfortable as they can be with a person stepping on them.

SMART FURNITURE

Adaptive structures will be useful in furniture. Today, we have furniture that adapts to the human body, but it does so in an awk-

ward and incomplete manner. It adapts because people grab cushions
and move them around. Or a chair adapts because it is a hinged
contraption that grudgingly bends and extends in a few places to suit
a small range of preferred positions. Occasionally, one sees furniture
that allegedly gives a massage, but in fact only vibrates.

These limitations are consequences of the expense, bulkiness,
clumsiness, and unreliability of such things as moving parts, motors,
sensors, and computers today. With molecular manufacturing, it will
be easy to make furniture from smart materials that can adapt to an
individual human body, and to a person's changing position, to consistently give comfortable support. Smart cushions could also do a
better job of responding to hints in the form of pats, tugs, and punches.
As for massage—a piece of furniture, no matter how advanced, is not
the same as a masseuse. Still, a typical massage setting on a smart
chair would not mean today's "vibrate medium vigorously," but
something closer to "five minutes of shiatsu."

AND SO FORTH . . .

This tour through of the potential of smart matter has shown how
we could get walls that look and sound as we wish, clothing, shoes,
and furniture of greater comfort, and clean solar power. As one might
expect, this just scratches the surface.

If you care to think of further applications, here are some ground
rules: Components made by molecular manufacturing can be many
tens of times stronger than steel, but materials made by plugging
many components together will be weaker. For these, strengths in
the range of cotton candy to steel seem achievable. The components
will be sensitive to heat, and at high temperatures they will break
down or burn. Many materials will be able to survive the temperature
of boiling water, but only specialized designs would be oven-safe.
Color, texture, and (usually) sound should be controllable. Surfaces
can be smooth and tightly sealed (this takes some cleverness). Motions can be fairly fast. Power has to come from somewhere; good
sources include electricity, stored chemical energy, and light. If na-

nomachines or smart materials are dunked in liquids, chemical energy can come from dissolved molecules; if they are in the open, energy can come from light; if they are sitting in one place, they can be plugged into a socket; if they are moving around in the dark, they can run on batteries for a while, then run down and quit. Within these limits, much can be accomplished.

"Smart" is a relative term. Unless you want to assume that people learn a lot more about intelligence and programming, it is best to assume that these materials will follow simple rules, like those followed by parts of drawings on computer screens. In these drawings, a picture of a rectangle can be commanded to sprout handles at its corners; pulling a handle stretches or shrinks the rectangle without distorting its right-angle corners. An object made of smart matter could do likewise in the real world: a box could be stretched to a different size, then made rigid again; a door in a smart-material wall could have its *position* unlocked, its frame moved a pace to the left, and then be returned to normal use.

There seems little reason to make bits of smart matter independent, self-replicating, or toxic. With care, smart matter should be safer than what it replaces because it will be better controlled. Spray paint gets all over things and contains noxious solvents; the paper-paint described above doesn't. This will be a characteristic difference, if we exercise our usual vigilance to encourage the production of things that are safe and environmentally sound.

FALLING COSTS

It may be fun to discuss wondrous new products, but they won't make much difference in the world if they are too expensive. Besides, many people today don't have decent food, clothes, and a roof over their heads, to say nothing of fancy "nanostuff."

Costs matter. There is more to life than material goods, but without material goods life is miserable and narrow. If goods are expensive, people strive for them; if goods are abundant, people can turn their attention elsewhere. Some of us like to think that we are above a concern for material goods, but this seems more common in the wealthy countries. Lowering manufacturing costs is a mundane concern, but so are feeding people, housing them, and building sewage systems to keep them from dying of cholera and hepatitis. For all these reasons, finding ways to bring down production costs is a worthy goal.

For the poor, for the environment, and for the freeing of human potential, costs matter deeply. Let's take a closer look at the costs of molecular manufacturing.

CAN FALLING COSTS BE REALISTIC?

Inflation produces the illusion that costs rise, when the real story is that the value of money is falling. In the short term, real costs usually don't change very quickly, and this can produce the illusion that costs are stable facts of nature, like the law of gravity or the laws of thermodynamics.

In the real world, though, most costs have been falling by a crucial measure: the amount of human labor needed to make things. People can afford more and more, because their labor, supplemented by machines, can produce more and more. This change is dramatic measured on a scale of centuries, and equally dramatic across the gulf between Third World and developed countries. The rise from Third World to First World standards of living has raised income (dropped the cost of labor time) by more than a factor of ten. What can molecular manufacturing do?

Larger cost reductions have happened, most dramatically in computers. The cost of a computer of a given ability has fallen by roughly a factor of 10 every seven years since the 1940s. In total, this is a factor of a *million*. If automotive technologies had done likewise, a luxury car would now cost less than one cent. (Personal computer systems still cost hundreds of dollars both because they are far more powerful than the giant machines of the 1940s and because the cost of buying any useful computer *system* includes much more than just the cost of a bare computer chip.)

COSTS: A FIRST ESTIMATE

Some costs apply to a *kind* of product, regardless of how many copies of it are made: these include design costs, technology-licensing costs, regulatory-approval costs, and the like. Other costs apply to each *unit* of a product: these include the costs of labor, energy, raw materials, production equipment, production sites, insurance, and waste disposal. The per-kind costs can become very low if production runs are large. If these costs stay high, it will be because people prefer new products for their new benefits, despite the cost—hardly cause for complaint.

The more basic and easier to analyze costs are per-unit costs. A picture to keep in mind here is of Desert Rose Industries, where molecular machinery does most of the work, and where products are made from parts that are ultimately made from simple chemical substances. Let's consider some cost components.

Energy: Manufacturing at the molecular scale need not use a lot of energy. Plants build billions of tons of highly patterned material every year using available solar energy. Molecular manufacturing can be efficient, in the sense that the energy needed to build a block of product should be comparable to the energy released in burning an equivalent mass of wood or coal. If this energy were supplied as electricity at today's costs, the energy cost of manufacturing would be something like a dollar per kilogram. We'll return to the cost of energy later.

Raw Materials: Molecular manufacturing won't need exotic materials as inputs. Plain bulk chemicals will suffice, and this means materials no more exotic than the fuels and feedstocks that are, for now, derived from petroleum and biomass—gasoline, methanol, ammonia, and hydrogen. These typically cost tens of cents per kilogram. If bizarre compounds are used, they can be made internally. Rare elements could be avoided, but might be useful in trace amounts. The total quantity of raw materials consumed will be smaller than in conventional manufacturing processes because less will be wasted.

Capital Equipment and Maintenance: As we saw in the Desert Rose scenario, molecular manufacturing can be used to build all of the equipment needed for molecular manufacturing. It seems that this equipment—everything from large vats to submicroscopic special-purpose assemblers—can be reasonably durable, lasting for months or years before being recycled and replaced. If the equipment were to cost dollars per kilogram, and produce many thousands of kilograms of product in its life, the cost of the equipment would add little to the cost of the product.

Waste Disposal: Today's manufacturing waste is dumped into the air, water, and landfills. There need be no such waste with molecular manufacturing. Excess materials of the kind now spewed into the environment could instead be completely recycled internally, or could emerge from the manufacturing process in pure form, ready for use in some other process. In an advanced process, the only wastes would be leftover atoms resulting from a bad mix of raw materials. Most of these leftover atoms would be ordinary minerals and simple gases like oxygen, the main "waste" from the molecular machinery of plants. Molecular manufacturing produces no new elements—if arsenic comes out, arsenic must have gone in, and the process isn't to blame for its existence. Any intrinsically toxic materials of this sort can at least be put in the safest form we can devise for disposal. One option would be to chemically bond it into a stable mineral and put it back where it came from.

Labor: Once a plant is operating, it should require little human labor (what people do with their time will change, unless factories are kept running as bizarre hobbies). Desert Rose Industries was run by two people, yet was described as producing large quantities of varied goods. The basic molecular-scale operations of manufacturing have to be automated, since they are too small for people to work on. The other operations are fairly simple and can be aided by equipment for handling materials and information.

Space: Even a manufacturing plant based on nanotechnology takes up room. It would, however, be more compact than familiar manufacturing plants, and could be built in some out-of-the-way place with inexpensive land. These costs should be small by today's standards.

Insurance: This cost will depend on the state of the law, but some comparisons can be made. Improved sensors and alarms could be made integral parts of products; these should lower fire and theft premiums. Product liability costs should be reduced by safer, more reliable products (we'll discuss the question of product safety further in Chapter 12). Employee injury rates will be reduced by having less labor input. Still, the legal system in the United States has shown a disturbing tendency to block every *new* risk, however small, even when this forces people to keep suffering *old* risks, which are sometimes huge. (The supply of lifesaving vaccines has been threatened in just this way.) When this happens, we kill anonymous people in the name of safety. If this behavior raises insurance premiums in a perverse way, it could discourage a shift to safer manufacturing technologies. Since such costs can grow or shrink independent of the real world of engineering and human welfare, they are beyond our ability to estimate.

Sales, Distribution, Training . . . : These costs will depend on the product: Is it as common as potatoes, and as simple to use? Or is it rare and complex, so that determining what you need, where to get it, and how to use it are the main problems? These service costs are real but can be distinguished from costs of the thing itself.

To summarize, molecular manufacturing should eventually lead to lower costs. The initial expense of developing the technology and specific products will be substantial, but the cost of production can be low. Energy costs (at present prices) and materials costs (ditto) would be significant, but not enormous. They were quoted on a per-kilogram basis, but nanotechnological products, being made of superior materials, will often weigh only a fraction of what familiar products do. (Ballast, were it needed, will be dirt-cheap.) Equipment costs, land costs, waste-disposal costs, and labor costs can be low by the very nature of the technology.

Costs of design, regulation, and insurance will depend strongly on human tastes and are beyond predicting. Basic products, like clothing and housing, can become inexpensive unless we do something to keep them costly. As the cost of improved safety falls, there will be less reason to accept unsafe products. Molecular manufacturing uses processes as controlled and efficient as the molecular processes in plants. Its products could be as inexpensive as potatoes. This may sound too good to be true (and there *are* downsides, as we'll discuss), but why shouldn't it be true? Shouldn't we expect large changes to come with the replacement of modern technology?

A Cycle of Falling Costs

The above estimate made a conservative assumption about future costs: that energy and materials will cost then what they do now, before molecular manufacturing has become available. They won't, because lower costs lead to lower costs.

Let's say that making one kilogram of product by molecular manufacturing requires one dollar a kilogram of raw materials and four dollars for a generous forty kilowatt-hours of energy. These are typical present-day prices for materials and electrical energy. Assume, for the moment, that other costs are small. One of the resulting five-dollars-per-kilogram products can be solar-cell paint suitable for applying to paved roads. A layer of paint a few millionths of a meter thick would cost about five cents per square meter to produce, and would generate enough energy to make another square meter of paint in less than a week, even allowing for nighttime and moderate cloud cover. The so-called energy payback time would thus be short.

Let's assume that this smart paint costs as much to spread and hook up as it does to make, and that we demand that it pay for itself in a single month, so we charge ten cents per square meter per month. At that rate, the cost of solar energy from resurfaced roads would be roughly $0.004 per kilowatt hour—less than a twentieth the energy cost assumed in the initial production-cost estimate. By itself, this makes the cost of production fall to a fraction of what it was before. Most of that remaining fraction consists of the cost of materials.

But the products of nanotechnology will mostly be made of carbon (if present expectations are any guide), and carbon dioxide is too abundant in the atmosphere these days. With energy so cheap, the atmosphere can be used as a source of carbon (and of hydrogen, nitrogen, and oxygen). The price of carbon would be a few cents per kilogram—roughly a twentieth the original price assumed for raw materials.

But now, both energy and raw materials are a twentieth the original price, and so the products become cheaper, including the energy-producing products and the raw-material–producing (atmosphere-cleaning) products. . . .

The above scenario is simple, but it seems realistic in its basic outlines: lower costs can lead to lower costs. How far this process can go is hard to estimate precisely, but it could go far indeed.

POWER TOO CHEAP TO METER?

This argument will remind some readers of an old claim—that nuclear energy would lead to "power too cheap to meter." This assertion, attributed to the early nuclear era, has passed into folklore as a warning to be skeptical of technologists promising free goodies. Does the warning apply here?

Anyone claiming that something is free doesn't really understand economics. Using something always has a cost equal to the most valuable alternative use for the thing. Choosing one alternative sacrifices another, and that sacrifice is the cost. As economist Phillip K. Salin says, "There's no such thing as a free opportunity," since opportunities always cost (at least) time and attention. Nanotechnology will *not* mean free goodies.

But, one might argue, nuclear power hasn't even been inexpensive. If technologists could be so wrong back then, why believe a similar argument today? We are happy to report that the arguments aren't similar: any argument for "nuclear power too cheap to meter" had to be absurd *even given the knowledge at the time*, and our argument isn't.

Nuclear reactors boil water to make steam to turn turbines to turn generators to drive electrical power through power lines to transformers to local power lines to houses, factories, and so forth. The wildest optimist could never have claimed that nuclear power was a free source of anything more than heat, and a realist would have added in the cost of the reactor equipment, fuel, waste disposal, hazards, and the rest. Even our wild optimist would have had to include the cost of building the boiler, the turbines, the generators, the power lines, and the transformers, and the cost of maintenance on all these. These costs were known to be a major part of the cost of power, so free heat wouldn't have meant free power. Thus, the claim was absurd the day it was made—not merely in hindsight.

In the early 1960s, Alvin Weinberg, head of the Oak Ridge National Laboratory, was a strong advocate of nuclear power, and argued that it would provide "cheap energy." He was optimistic, but did his sums. First, he assumed that nuclear-power plants could be built a little more cheaply than coal-fired power plants of the same size. Then he assumed that the cost of fuel, waste disposal, operations, and maintenance for nuclear plants would be not much more than the cost of operations and maintenance alone for coal plants. Then he assumed that they might last for more than thirty years. Finally, he assumed that they would be publicly operated, tax free at low interest (which merely moves costs elsewhere) and that after thirty years the cost of the equipment would be written off (which is an accounting fiction). With all of that, he derived a power cost that "might be" as low as *one half* the cost of the cheapest coal-fired plant he mentions. He was clearly an optimist, but he didn't come close to arguing for power too cheap to meter.

Low but Not Zero Costs

People have cried "Wolf!" before about new technologies leading to overwhelming abundance. It was said of nuclear power, and of steam power before it, and perhaps of water wheels, the horse, the plough, and the chipped rock. Molecular manufacturing is different because it is a new way to make almost anything, including more of the equipment needed to do the manufacturing. There has never been anything quite like this before.

The basic argument for low-cost production is this: Molecular manufacturing will be able to make almost anything with little labor, land, or maintenance, with high productivity, and with modest requirements for materials and energy. Its products will themselves be extremely productive, as energy producers, as materials collectors, and as manufacturing equipment. There has never been a technology with this combination of characteristics, so historical analogies must be used with care. Perhaps the best analogy is this: Molecular manufacturing will do for matter processing what the computer has done for information processing.

There will always be limiting costs, because resources—whether energy, matter, or design skill—always have some alternative use. Costs will not fall to zero, but it seems that they could fall very low indeed.

EIGHT

Providing the Basics, and More

The hungry, the homeless, and the hunted have little time or energy to devote to human relations or personal development. Food, shelter, and security are not everything, but they are basic. Material abundance is perhaps the best-known way to build a contempt for material things and a concern for what lies beyond. In that spirit, let us take a further look at providing heaps of basic material wealth where today there is poverty.

The idea of bringing everyone in the world up to a decent standard of living looks utopian today. The world's poor are numerous and the wealthy are few, and yet the Earth's resources are already strained by our crude industrial and agricultural technologies. For the 1970s and 1980s, with a growing awareness of the environmental impact of human population and pollution, many people have begun to wrestle with the specter of declining wealth. Few have allowed themselves to consider what it might be like to live in a world with far greater material wealth because it has seemed impossible. Any discussion of such things will inevitably have a whiff of the 1950s or 1960s about it: Gee whiz, we can have supercars and Better Living

Through (a substitute for conventional) Chemistry!

In the long run, unless population growth is limited, it will be impossible to maintain a decent standard of living for everyone. This is a basic fact, and to ignore it would be to destroy our future. Yet within sight is a time in which the world's poorest can be raised to a material standard of living that would be the envy of the world's richest today. The key is efficient, low-cost production of high-quality goods. Whether this will be used to achieve the goals we describe is more than just a question of technology.

Here, as in the next two chapters, we continue to focus on how the new technologies can serve positive goals. There is a lot to say, and it needs to be said, in part because positive goals can in some measure displace negative goals. We ask patience of those readers bothered by what may seem an optimistic tone, and ask that they imagine the authors to share their fears that powerful technologies will be abused, that positive goals may end in ruin, that a material paradise may yet harbor human misery. Chapters 11 and 12 will discuss limits, accidents, and abuse.

THIRD WORLD NANOTECHNOLOGY

Where wealth is concerned, the least-developed countries present the hardest case. Can a capability as advanced as nanotechnology, based on molecular machinery, be of use in the Third World? The answer must be yes. Agriculture is the backbone of Third World economies today, and agriculture is based on the naturally occurring molecular machines in wheat, rice, yams, and the like.

The Third World is short on equipment and skills. (It often has governmental problems as well, but that is another story.) Molecular manufacturing can make equipment inexpensive enough for the poor to buy or for aid agencies to give away. This includes equipment for making more equipment, so dependency could be reduced. As for skills, basic molecular manufacturing will require little labor of any kind, and a little skill will go a long way. As the technology advances, more and more of the products can be easy-to-use smart materials.

Molecular manufacturing will enable the poorest countries to by-pass the difficult and dirty process of the industrial revolution. It can make products that are less expensive and easier to use than yams or rice or goats or water buffalo. And with products like cheap super-computers with huge databases of writing and animation viewed through 3-D color displays, it can even help spread knowledge.

Nanotechnology's role in helping the poorest nations won't be on the minds of the first developers—they'll be in government and commercial labs in the wealthiest nations, pursuing problems of concern to people there. History, though, is full of unintended consequences, and some are for the better.

CONSTRUCTION AND HOUSING

Building large objects is basic to solving problems of housing and transportation. Smart materials can help.

Today, buildings are expensive to construct, expensive to replace, and expensive to make fireproof, tornado-proof, earthquake-proof, and so forth. Making buildings tall is expensive; making walls soundproof is expensive; building underground is expensive. Efforts to relieve city congestion often founder on the high cost of building subways, which can amount to hundreds of millions of dollars per mile.

Building codes and politics permitting, nanotechnology will make possible revolutions in the construction of buildings. Superior materials will make it easy to construct tall (or deep) buildings to free up land, and strong buildings that can ride out the greatest earthquake without harm. Buildings can be made so energy-efficient and so good at using the solar energy falling on them that most are net energy producers. What is more, smart materials can make it easy to build and modify complex structures, such as walls full of windows, wiring, plumbing, data networks, and the like. For a concrete example that shows the principle, let's picture what smart pipes could be like.

Let's say that you want to install a fold-down sink in the corner of your bedroom. The new materials make fold-down sinks practical, and in a house made of advanced smart materials, just sticking one on the wall would be enough—the plumbing would rearrange itself. But this is an old, pre-breakthrough house, so the sink is a retrofit. To do this home-handiwork project, you buy several boxes full of inexpensive tubing, T-joints, valves, and fixtures in a variety of sizes, all as light as wood veneer and feeling like soft rubber.

The biggest practical problem will be to make a hole from an existing water pipe and drainpipe to where you want the sink. Molecular manufacturing can provide excellent power tools to make the holes, and smart paint and plaster to cover them again, but the details depend on how your house is built.

The smart plumbing system does help, of course. If you want to run the drain line through the attic, built-in pumps will make sure that the water flows properly. The flexibility of the pipes makes it much easier to run them around curves and corners. Low-cost power makes it practical for the sink to have a flow-through water heater, so you only need to run a cold-water pipe to have both hot and cold water. All the parts go together as easily as a child's blocks, and seem about as flimsy and likely to leak. When you turn it on, though, the microscopic components of the pipes lock together and become as strong as steel. And smart plumbing *doesn't* leak.

If your house were made of smart materials, like most of the housing in the Third World these days, life would have been easier. Using a special trowel, wall structures would be reworked like soft clay, doing their structural job all the while. Setting up a plumbing system from scratch with this stuff is easy, and hard to do wrong. Drinking-water pipes won't connect to wastewater pipes, so drinking water can't be accidentally contaminated. Drains won't clog, because they can clean themselves better than a rotary steel blade ever could. If you run enough pipes from everything to everything else, built-in pumps will make sure that water flows in the right direction with adequate pressure.

Smart plumbing is one example of a general pattern. Molecular manufacturing can eventually make complex products at low cost, and those complex products can be simpler to use than anything we have today, freeing our attention for other concerns. Buildings can become easy to make and easy to change. The basic conveniences of the modern world, and more, can be carried to the ends of the earth and installed by the people there to suit their tastes.

FOOD

Worldwide food production has been outpacing population growth, yet hunger continues. In recent years, famine has often had political roots, as in Ethiopia where the rulers aim to starve opponents into submission. Such problems are beyond a simple technological solution. To avoid getting headaches, we'll also ignore the politics of farm price-support programs, which raise food prices while people are going hungry. All we can suggest here is a way to provide fresh food at lower cost with reduced environmental impact.

For decades, futurists have predicted the coming of synthetic foods. Some sort of molecular-manufacturing process could doubtless make such things with the usual low costs, but this doesn't sound appetizing, so we'll ignore the idea.

Most agriculture today is inefficient—an environmental disaster. Modern agriculture is famed for wasting water and polluting it with synthetic fertilizers, and for spreading herbicides and pesticides over the landscape. Yet the greatest environmental impact of agriculture is its sheer consumption of land. In the American East, ancient forests disappeared under the ax, in part to supply wood, in part to clear land. The prairies of the West disappeared under the plow. Around the world, this trend continues. The technology of the ax, the fire, and the plow is chiefly responsible for the destruction of rain forests today. A growing population will tend to turn every productive ecosystem into some sort of farmland or grazing land, if we let it.

No technological fix can solve the long-term problem of population growth. Nonetheless, we can roll back the problem of the loss of land, yet increase food supplies. One approach is intensive greenhouse agriculture.

Every kind of plant has its optimum growing conditions, and those conditions are far different from those found in most farmland during most of the year. Plants growing outdoors face insect pests, unless doused with pesticide, and low levels of nutrients, unless doused with fertilizer. In greenhouses patrolled by "nanoflyswatters" able to eliminate invading insects, plants would be protected from pests and could be provided with nutrients without contaminating groundwater or runoff. Most plants prefer higher humidity than most climates provide. Most plants prefer higher, more uniform temperatures than are typically found outdoors. What is more, plants thrive in high levels of carbon dioxide. Only greenhouses can provide pest protection, ample nutrients, humidity, warmth, and carbon dioxide all together and without reengineering the Earth.

Taken together, these factors make a *huge* difference in agricultural productivity. Experiments with intensive greenhouse agriculture, performed by the Environmental Research Lab in Arizona, show that an area of 250 square meters—about the size of a tennis court—can raise enough food for one person, year in and year out. With molecular manufacturing to make inexpensive, reliable equipment, the intensive labor of intensive agriculture can be automated. With technology like the deployable "tents" and smart materials we have described, greenhouse construction can be inexpensive. Following the standard argument, with equipment costs, labor costs, materials costs, and so forth all expected to be low, greenhouse-grown foods can be inexpensive.

What does this mean for the environment? It means that the human race could feed itself with ordinary, naturally grown, pesticide-free foods while returning more than 90 percent of today's agricultural land to wilds. With a generous five-hundred square meters per person, the U.S. population would require only 3 percent of present U.S. farm acreage, freeing 97 percent for other uses, or for a gradual return to wilderness. When farmers are able to grow high-quality foodstuffs inexpensively, in a fraction of the room that they require today, they will find more demand for their land to be tended as a park or wilderness than as a cornfield. Farm journals can be expected to carry articles advising on techniques for rapid and aesthetic restoration of forest and grassland, and on how best to accommodate the desires of the discriminating nature lover and conservationist. Even "unpopular" land will tend to become popular with people seeking solitude.

The economics of assembler-based manufacturing will remove the incentive to make greenhouses cheap, ugly, and boxy; the only reason to build that way today is the high cost of building anything at all. And while today's greenhouses suffer from viral and fungal infestations, these could be eradicated from plants in the same way they would be from the human body, as will be described later. A problem faced by today's greenhouses—overheating—could be dealt with by using heat exchangers, thereby conserving the carefully balanced inside atmosphere. Finally, if it should turn out that a little bit of bad weather improves the taste of tomatoes, that, too, could be provided, since there would be no reason to be fanatical about sheer efficiency.

COMMUNICATIONS

Today, telecommunications systems have sharply limited capacity and are expensive to expand. Molecular manufacturing will drop the price of the "boxes" in telecommunications systems—things such as switching systems, computers, telephones, and even the fabled videophone. Cables made of smart materials can make these devices easy to install and easy to connect together.

Regulatory agencies willing, you might someday be able to buy inexpensive spools of material resembling kite string, and other spools of material resembling tape, then use them to join a world data network. Either kind of strand can configure its core into a good-quality optical fiber, with special provisions for going around bends. When rubbed together, pieces of string will fuse together, or fuse to a piece of tape. Pieces of tape do likewise. To hook up to the network, you run string or tape from your telephone or other data terminal to the nearest point that is already connected. If you live deep in a tropical rain forest, run a string to the village satellite link.

These data-cable materials include amplifiers, nanocomputers, switching nodes, and the rest, and they come loaded with software that "knows" how to act to transmit data reliably. If you're worried that a line may break, run three in different directions. Even one line could carry far more data than all the channels in a television cable put together.

TRANSPORTATION

Getting around quickly requires vehicles and somewhere for them to travel. The old 1950s vision of private helicopters would be technically possible with inexpensive, high-quality manufacturing, cheap energy, and a bit of improvement in autopilots and air-traffic control—but will people really tolerate that much junk roaring across the sky? Fortunately, there is an alternative both to this and to building ever-more roads.

GOING UNDERGROUND

Near the surface of the Earth, there is as much room underground as there is above it. This is usually ignored, because the room is full of dirt, rock, pressurized water, and the like. Digging is expensive. Digging long, deep tunnels is even more expensive. This expense, however, is mostly in the cost of equipment, materials, and energy. Tunneling machines are in wide use today, and molecular manufacturing can make them more efficient and less expensive. The energy to operate them will be no great problem, and smart materials can line tunnels as fast as they are dug, with little or no labor. Nanotechnology will open the low frontier.

With a little care, the environmental impact of a deep tunnel can be trivial. Instead of solid rock far below the surface, there is rock with a sealed tunnel running through it. Nothing nearby need be disturbed.

Tunnels avoid both the aesthetic impact of a sky full of noisy aircraft and the environmental impact of paving strips of landscape. This will make them less expensive than roads, and they can, if desired, be more common than roads in the developed world today. They will even permit faster transportation.

TAKING THE SUBWAY

Japan and Germany are actively developing magnetic trains, like those in the Desert Rose scenario. These avoid the limitations of steel wheels on steel rails by using magnetic forces to "fly" the train along a special track. Magnetic trains can reach aircraft speeds at ground level. On long runs through evacuated tunnels, they can reach spacecraft speeds, traveling global distances in an hour or so (less, if passengers are willing to tolerate substantial acceleration).

Systems like this can give "taking the subway" a new meaning. Local transportation would be at fast automotive speeds, but long-distance transportation would be faster than the Concorde. With superconducting electrical systems, fast subways would be more energy-efficient than today's slow mass transit.

GETTING YOUR CAR

For decades, people have proposed replacing automobiles with some form of mass-transportation system, and it seems that cost revolutions (including inexpensive tunneling) may finally make this practical. Before junking the car, though, it's worth seeing how it might be improved.

Molecular manufacturing can make almost anything better. Automobiles can be made stronger and safer, lighter, higher performance, and higher efficiency, while getting excellent mileage and burning clean, inexpensive fuels, perhaps in fuel cells powering quiet electric motors. Using aerodynamic forces to hold the car to the road, there's no reason why a comfortable passenger car shouldn't be able to deliver uncomfortable, drag-racer acceleration.

To imagine a cheap car built with molecular manufacturing, first imagine loading it with all the attractive features that you've ever heard proposed. This includes everything from today's self-adjusting seats and mirrors, excellent sound systems, and specially tuned steering and suspension systems, through automated navigation displays, emergency braking, and reliable super-duper airbags. Now, instead of just having the position of the seats, mirrors, and so forth adjust to a driver, as some cars do today, our smart-material car can also adjust its size, shape, and color, facing owners with choices such as, "What should our car look like for this occasion?"

Those seeking an image of solid conservatism and wealth won't drive such cheap cars; they will risk their necks in a certified antique car, made from the traditional steel, paint, and rubber. If environmental regulations permit it, the car might even have a genuine gasoline-burning engine. The latter can no doubt be cleaned up by fancy nanotechnology-based emission-control systems.

OPENING THE SPACE FRONTIER

Our transportation system today effectively ends in the upper atmosphere. Travel beyond still takes the form of "historic missions." There is no reason for this situation to continue for long, once molecular manufacturing becomes well established.

The cost of spaceflight is high because spacecraft are huge, fragile things, made in such small numbers that they're almost hand-crafted. Molecular manufacturing will replace today's delicate monsters with rugged, mass-produced vehicles (which, with greater efficiency, needn't be so large). The vehicles will cost little, but the energy? Today, the energy cost of a ticket to orbit in an efficient vehicle would be less than one hundred dollars. Low-cost vehicles and energy will drop the total cost to a fraction of this.

We will know that spaceflight has become inexpensive when people see the Earth as just a small part of the world, and understand in their bones that space resources make continued exploitation of Earth's resources unnecessary. In the long run, efficient, clean, low-cost manufacturing can transform the way human beings affect the Earth by their presence. Even stay-at-home humans will be better able to heal the damage they have done.

NINE

Restoring the Environment

The 1970s saw a revolution in Western attitudes toward the natural environment. Concern with pollution, deforestation, and species extinction exploded. With the rise of these concerns came an ambivalent attitude toward technology and the wealth it was producing: some said that human beings are destructive to the environment in direct proportion to their power. This immediately suggested that technology and higher living standards were bad, being inherently destructive. "Wealth" came to imply environmental destruction.

The revolution in attitudes toward the environment has changed the idea of wealth. Our national statistics may not reflect it—not every last citizen or politician may agree—but the concept that genuine wealth includes not just houses and refrigerators, factories and machines, cars and roads, but also fields and forests, owls and wolves, clean air, clean water, and wilderness has taken deep root in minds and in politics. "The wealth of nature" has come to include nature as a value in itself, not merely as potential lumber, ore, and farmland.

As a consequence, greater wealth has begun to mean cleaner

wealth, greener wealth. Richer countries can afford more expensive, more efficient equipment—scrubbers on smokestacks, catalytic converters on cars—and so they can produce goods with less environmental impact. This trend gives at best a hint of the future.

Lester Milbrath, director of the Research Program in Environment and Society at the State University of New York at Buffalo, observes, "Nanotechnologies have the potential to produce plentiful consumer goods with much lower throughput of materials and much less production of waste, thus reducing carbon dioxide buildup and reducing global warming. They also have the potential to reduce waste, especially hazardous waste, converting it to natural materials which do not threaten life." James Lovelock states, "The future could be good if we regain a sense of purpose and embrace the new industries based on information and nanotechnology. These add enormous value to molecular-sized pieces of matter, and need not be a threat to the environment as were the heavy polluting industries of the past."

MAKING IT EASIER TO BE CLEAN

Should we boast of "high technology" while industry still can't produce without polluting? Pollution is a sign of low technology, of inadequate control of how matter is handled. Inferior goods and hazardous wastes are two sides of one problem.

With processes based on molecular manufacturing, industries will produce superior goods, and by virtue of the same advance in control, will have no need of burning, oiling, washing with solvents and acids, and flushing noxious chemicals down their drains. Molecular-manufacturing processes will rearrange atoms in controlled ways, and can neatly package any unwanted atoms for recycling or return to their source. This intrinsic cleanliness inspired environmentalist Terence McKenna, writing in the *Whole Earth Review*, to call nanotechnology "the most radical of the green visions."

This green vision will not be fulfilled automatically, but only with effort. Any powerful technology can be used for good or ill, and nanotechnology is no exception. Today, we see scattered progress in environmental cleanup and restoration, some slowing of ecological destruction, because of organized political pressure buoyed by a groundswell of public concern. Yet for all its force, this pressure is spread desperately thin, fighting enormous resistance rooted in economic forces.

But if these economic forces vanish, the opposition will crumble. Often, the key to success in battle is to give one's opponents an attractive alternative to fighting. The most powerful cry of the anti-green opposition has been that clearing and polluting the land offer the only path to wealth, the only escape from poverty. Now we can see a clean, efficient, and unobtrusive alternative: green wealth, compatible with natural wealth.

ENDING CHEMICAL POLLUTION, CUTTING RESOURCE CONSUMPTION

We've already seen how molecular manufacturing can provide clean solar energy without paving over desert ecosystems, and how clean energy and common materials can be turned into abundant, efficient goods, also cleanly. With care, sources of chemical pollution—even of excess carbon dioxide—can, step by step, be eliminated. This includes the pollutants responsible for acid rain, as well as ozone-destroying gases, greenhouse gases, oil spills, and toxic wastes.

In each case, the story is about the same. Acid rain mostly results from burning dirty fuels containing sulfur, and from burning cleaner fuels in a dirty way, producing nitrogen oxides. We've seen how molecular manufacturing can make solar cells cheap enough and rugged enough to use as road surfaces. With green wealth, we can make clean fuels from solar energy, air, and water; consuming these fuels in clean nanomechanical systems would just return to the air exactly the materials taken from it, along with a little water vapor. Fuels are made, fuels are consumed, and the cycle produces no net pollution. With cheap solar fuels, coal and petroleum can be replaced, ignored, left in the ground. When petroleum is obsolete, oil spills will vanish.

The greenhouse gas of greatest concern is carbon dioxide, and its main source is the burning of fossil fuels. The above steps would end this. The release of other gases, such as the chlorofluorocarbons (CFCs) used in foaming plastics, is often a side effect of primitive manufacturing processes: *foaming plastic* will hardly be a popular activity in an era of molecular manufacturing. These materials can be replaced or controlled—and they include the gases most responsible for ozone depletion.

The chief threats to the ozone layer are those same CFCs, used as refrigerants and solvents. Molecular manufacturing will use solvents sparingly (mostly water), and can recycle them without dumping any. CFC refrigerants can be replaced even with current technology, at a cost; with nanotechnology, that cost will be negligible.

Toxic wastes generally consist of harmless atoms arranged into noxious molecules; the same is true of sewage. With inexpensive energy and equipment able to work at the molecular level, these wastes can be converted into harmless forms. Many need never be produced in the first place. Other toxic wastes contain toxic elements, such as lead, mercury, arsenic, and cadmium. These elements come from the ground, and are best returned to the location and condition in which they were found. With nanotechnology, moreover, there will be little reason to dig them up in the first place. Nanotechnology will be able to break materials down to simple molecules and build them back up again. Need it be said that this will permit complete recycling?

It is fair to say that eliminating these sources of pollution would be a major improvement. There doesn't seem to be much more to say, aside from the usual caveats: "Not immediately," "Not all at once," and "Not on a predictable schedule." No one wants to make and dump wastes; they want something else, and get wastes as by-products. With a better way to get what people want, dumping wastes can be stopped.

People will also be able to get what they want while reducing their resource consumption. As materials grow stronger, they can be used more sparingly. As machines grow more perfect—in their motors, bearings, insulation, computers—they will grow more efficient.

Materials will be needed to make things, and energy will be needed to run them, but in smaller amounts. What is more, nanotechnology will be the ultimate recycling technology. Objects can be made extremely durable, decreasing the need for recycling; alternatively, objects can be made genuinely biodegradable, designed at the molecular level to decompose after use, leaving humus and mineral grit; alternatively, they can be made of microscopic snap-together pieces, making objects as recyclable as structures built and rebuilt out of a child's blocks; finally, even objects not designed for recycling can be taken apart into simple molecules and recycled regardless. Each approach has different advantages and costs, and each makes current garbage problems go away.

CLEANING UP THE TWENTIETH-CENTURY MESS

Still, even after twentieth-century industry is history, its toxic residues will remain. Cleaning up waste dumps with today's technology has proved so expensive and ineffective that many in the field have all but given up hope of really solving the problem. What can be done with post-breakthrough technologies?

CLEANSING SOIL AND WATER

Nanotechnology can help with the cleanup of these pollutants. Living organisms clean the environment, when they can, by using molecular machinery to break down toxic materials. Systems built with nanotechnology will be able to do likewise, and to deal with compounds that aren't biodegradable.

Alan Liss is director of research for Ecological Engineering Associates, a company that uses knowledge of how natural ecosystems function to address environmental problems such as wastewater treatment. He explains how cleanup could work: "The more we learn about the ecosystem, the more we find that functions are managed by particular organisms or groups of organisms. Nanotech 'managers' might be able to step in when the natural managers are not available, thereby having a particular ecological activity occur that otherwise wouldn't have happened. A nanotech manager might be used for remediation in a situation where toxicants have destroyed some key members of a particular ecosystem—some managerial microbes, for example. Once the needed activities are reinitiated, the living survivors of the stressed ecosystem can jump in and continue the ecosystem recovery effort."

To see how nanomachines could be used to clean up pollution, imagine a device made of smart materials and roughly resembling a tree, once it has been delivered and unfolded. Above ground are solar-collecting panels; below ground, a branching system of rootlike tubes reaches a certain distance into the soil. By extending into a toxic-waste dump, these rootlike structures could soak up toxic chemicals, using energy from the solar collectors to convert them into harmless compounds. Rootlike structures extending down into the water table could do the same cleanup job in polluted aquifers.

CLEANSING THE ATMOSPHERE

Most atmospheric pollutants are quickly washed out by rain (turning them into soil- and water-pollution problems), but some air pollutants are longer lasting. Among these are the chlorine compounds attacking the ozone layer that protects the Earth from excessive ultraviolet radiation. Since 1975, observers have recorded growing holes in the ozone layer: at the South Pole, the hole can reach as far as the tips of South America, Africa, and Australia. Loss of this protection subjects people to an increased risk of skin cancer and has unknown effects on ecosystems. The new technology base will be able to stop the increase in ozone-destroying compounds, but the effects would linger for years. How might this problem be reversed more rapidly?

Thus far, we've talked about nanotechnology in the laboratory, in manufacturing plants, and in products for direct human use. Mo-

FIGURE 10: ENVIRONMENTAL CLEANUP

By changing the way materials and products are made, molecular manufacturing will free up land formerly used for industrial plants. Toxic materials could be removed from contaminated soil using solar power as the energy source, and the cleanup device and any collected residues could later be carted away.

lecular manufacturing can also make products that will perform some useful temporary function when tossed out into the environment. Getting rid of ozone-destroying pollutants high in the stratosphere is one example. There may be simpler approaches, without the sophistication of nanotechnology, but here is one that would work to cleanse the stratosphere of chlorine: Make huge numbers of balloons, each the size of a grain of pollen and light enough to float up into the ozone layer. In each, place a small solar-power plant, a molecular-processing plant, and a microscopic grain of sodium. The processing plant collects chlorine-containing compounds and separates out the chlorine. Combining this with the sodium makes sodium chloride— ordinary salt. When the sodium is gone, the balloon collapses and falls. Eventually, a grain of salt and a biodegradable speck fall to Earth, usually at sea. The stratosphere is soon clean.

A larger problem (with a ground-based solution) is climatic change caused by rising carbon dioxide (CO_2) levels. Global warming, expected by most climatologists and probably under way today, is caused by changes in the composition of Earth's atmosphere. The sun shines on the Earth, warming it. The Earth radiates heat back into space, cooling. The rate at which it cools depends on how transparent the atmosphere is to the radiation of heat. The tendency of the atmosphere to hold heat, to block thermal radiation from escaping into space, causes what is called the "greenhouse effect." Several gases contribute to this, but CO_2 presents the most massive problem. Fossil fuels and deforestation both contribute. Before the new technology base arrives, something like 300 billion tons of excess CO_2 will likely have been added to the atmosphere.

Small greenhouses can help reverse the global greenhouse effect. By permitting more efficient agriculture, molecular manufacturing can free land for reforestation, helping to repair the devastation wrought by hungry people. Growing forests absorb CO_2.

If reforestation is not fast enough, inexpensive solar energy can be applied to remove CO_2 directly, producing oxygen and glossy graphite pebbles. Painting the world's roads with solar cells would yield about four trillion watts of power, enough to remove CO_2 at a rate of 10 billion tons per year. Temporarily planting one-tenth of

U.S. farm acreage with a solar-cell "crop" would provide enough energy to remove 300 billion tons in five years; winds would distribute the benefits worldwide. The twentieth-century insult to Earth's atmosphere can be reversed by less than a decade of twenty-first century repair work. Ecosystems damaged in the meantime are another matter.

ORBITAL WASTE

The space near Earth is being polluted with small orbiting projectiles, some as small as a pin. Most of the debris is floating fragments of discarded rocket stages, but it also includes gloves and cameras dropped by astronauts. This is not a problem for life on Earth, but it is a problem as life begins its historic spread beyond Earth—the first great expansion since the greening of the continents, long ago.

Orbiting objects travel much faster than rifle bullets, and energy increases as the square of speed. Small fragments of debris in space can do tremendous damage to a spacecraft, and worse—their impact on a spacecraft can blast loose yet more debris. Each fragment is potentially deadly to a spacefaring human crossing its path. Today, the tiny fraction of space that is near Earth is increasingly cluttered.

This litter needs to be picked up. With molecular manufacturing, it will be possible to build small spacecraft able to maneuver from orbit to orbit in space, picking up one piece of debris after another. Small spacecraft are needed, since it makes no sense to send a shuttle after a scrap of metal the size of a postage stamp. With these devices, we can clean the skies and keep them hospitable to life.

NUCLEAR WASTE

We've spoken of waste that just needs molecular changes to make it harmless, and toxic elements that came from the ground, but nuclear technology has created a third kind of waste. It has converted the slow, mild radioactivity of uranium into the fast, intense radioactivity of newly created nuclei, the products of fission and neutron bombardment. No molecular change can make them harmless, and these materials did not come from the ground. The products of molecular manufacturing could help with conventional approaches to dealing with nuclear waste, helping to store it in the most stable, reliable forms possible—but there is a more radical solution.

Even before the era of the nuclear reactor and the nuclear bomb, experimenters made artificially radioactive elements by accelerating particles and slamming them into nonradioactive targets. These particles traveled fast enough to penetrate the interior of an atom and reach the nucleus, joining it or breaking it apart.

The entire Earth is made of fallout from nuclear reactions in ancient stars. Its radioactivity is low because so much time has passed—many half-lives, for most radioactive nuclei. "Kicking" these stable nuclei changes them, often into a radioactive state. But kicking a radioactive nucleus has a certain chance of turning it into a stable one, destroying the radioactivity. By kicking, sorting, and kicking again, an atom-smashing machine could take in electrical power and radioactive waste, and output nothing but stable, nonradioactive elements, identical to those common in nature. Don't recommend this to your congressman—it would be far too expensive, today—but it will some day be practical to destroy the radioactivity of the twentieth-century's leftover nuclear waste.

Nanotechnology cannot do this directly, because molecular machines work with molecules, not nuclei. But *indirectly*, by making energy and equipment inexpensive, molecular manufacturing can give us the means for a clean, permanent solution to the problem of wastes left over from the nuclear era.

A WEALTH OF GARBAGE

Shortages often spur environmental damage. Faced with a food shortage, herdsmen can graze grasslands down to bare dirt. Faced with an energy shortage, industrial countries can approve destructive projects.

The growth of population and the consumption of resources by twentieth-century industry have placed growing pressures on Earth's ability to support us in the manner to which we have become accustomed.

The resource problem will look quite different in the twenty-first century, with a new technology base. Today, we cut trees and mine iron for our structures. We pump oil and mine coal for our energy. Even cement is born in the flames of burning fossil fuels. Almost everything we build, almost every move we make, consumes something ripped from the Earth. This need not continue.

Our civilization uses materials for many things, but mainly to make things with a certain size, shape, and strength. These structural uses include everything from fibers in clothing to paving in roads, and most of the mass of furniture, walls, cars, spacecraft, computers—indeed, most of the mass of almost every product we build and use. The best structural materials use carbon, in forms like diamond and graphite. With elements from air and water, carbon makes up the polymers of wool and polyester, and of wood and nylon. A twenty-first century civilization could mine the atmosphere for carbon, extracting over 300 billion tons before lowering the CO_2 concentration back to its natural, pre-industrial level. For a population of 10 billion, this would be enough to give every family a large house with lightweight but steel-strong walls, with 95 percent left over. Atmospheric garbage is an ample source of structural materials, with no need to cut trees or dig iron ore.

Plants show that carbon can be used to build solar collectors. Laboratory work shows that carbon compounds can be better conductors than copper. A whole power system could be built without even touching the rich resources of metal buried in garbage dumps.

Carbon can make windows, of plastic or diamond. Carbon can make things colorful with organic dyes. Carbon can be used to build nanocomputers, and will be the chief component of high-performance nanomachines of all kinds. The other components in all these materials are hydrogen, nitrogen, and oxygen, all found in air and water. Other elements are useful, but seldom necessary. Traces would often be ample.

With a new technology base making recycling easy, there need be no steady depletion of Earth's resources, just to keep a civilization running. The sketch just made shows that recycling just one form of garbage—excess atmospheric CO_2—can provide most needs. Even 10 billion wealthy people would not need to strip the Earth of resources. They could make do with what we've already dug up and thrown away, and they wouldn't even need all of that.

In short, a twenty-first-century civilization with a population of 10 billion could maintain a high standard of living using nothing but waste from twentieth-century industry, supplemented with modest amounts of air, water, and sunlight. This won't necessarily happen, yet the very fact that it is *possible* gives a better sense of what the new technology base can mean for the relationship between humanity, resources, and the Earth.

GREEN PRODUCTS

In *The Green Consumer*, Elkington, Hailes, and Makower define a green product as one that:

- Is not dangerous to the health of people or animals
- Does not cause damage to the environment during manufacture, use, or disposal
- Does not consume a disproportionate amount of energy and other resources during manufacture, use, or disposal
- Does not cause unnecessary waste, due either to excessive packaging or to a short useful life
- Does not involve the unnecessary use of or cruelty to animals
- Does not use materials derived from threatened species or environments
- Ideally, does not trade price, quality, nutrition, or convenience for environmental quality

With its ability to make almost anything at low cost—including products designed for extreme safety, durability, efficiency—without mining, logging, harming animals or environments, or producing toxic wastes, molecular manufacturing will make possible greener products than any yet seen in a store. Nanotechnology can replace dirty wealth with green wealth.

ENVIRONMENTAL RESTORATION

A central problem in environmental restoration is reversing environmental encroachment. We tend to see land as being gobbled up by housing, because the land *where we live* generally is. Farming, though, consumes more land, and the variant of farming called "forestry" consumes still more. By rolling back our requirement for farmland, and for wood and paper, nanotechnology can change the balance of forces behind environmental encroachment. This should make it more practical, politically and economically, for people to move toward environmental restoration.

Restoring the environment means returning land to what it was—removing what has been added and, where possible, replacing what has been lost. We've seen how this can be done, in part, by removing pollutants and some of the pressures for ploughing and paving. A more difficult problem, though, is restoring the ecological balance where the changes have been biological. Much of Earth's biological diversity has been a result of biological isolation, of islands, seas, mountains, and continents. This isolation has been breached, and reversing the resulting problems is one of the greatest challenges in healing the biosphere.

IMPORTED SPECIES

Human meddling with life in the biosphere has caused enormous ecological disruptions. This hasn't involved genetic engineering—by twisting organisms to better serve human purposes, genetic engineering usually leaves them less able to serve their own purposes, less able to survive and reproduce in the wild. The great disruptions have come from a different source: from globe-traveling human beings taking aggressive, well-adapted species from one part of the planet to another, landing them on a distant island or continent to invade an ecosystem with no evolved defenses. This has happened again and again.

Australia is a classic case. It had been isolated long enough to evolve its own peculiar species quite unfamiliar elsewhere: kangaroos, koalas, duckbilled platypuses. When humans arrived, they brought new species. Whoever brought the first rabbits could not have guessed that they, of all creatures, would be so destructive. They soon overran the continent, destroying crops and grazing lands, unchecked by natural competitors or predators. They were joined by invaders from the plant kingdom: the prickly pear and others.

The Americas have suffered invasions, too: tumbleweed, a bane of the rancher and farmer, is a relatively recent import from Central Asia. Since 1956, Africanized bees have been spreading from Brazil and moving north—but what they displace, in America, are European bees. Africa, in turn, is being invaded by the American screw-worm fly, an insect with larvae that enter an animal's wounds, including the umbilical wound of a newborn, and eat it alive. The story goes on and on.

People have sometimes tried, with a measure of success, to fight fire with fire: to bring in parasitic species and diseases to attack the imported species and keep its growth within some reasonable bounds. Australia's problem with prickly pear was tackled using an insect from Argentina; the rabbits were cut back—with mixed results—using a viral disease called myxomatosis: "rabbit pox."

Ecosystem Protectors

In many parts of the world, native species have been driven to extinction by rats, pigs, and other imported species, and others are endangered and fighting for their lives. Biological controls—fighting fire with fire—have advantages: organisms are small, selective, and inexpensive. These advantages will eventually be shared by devices made using molecular manufacturing, which avoid the disadvantages of importing and releasing yet more uncontrollable, breeding, spreading species. Alan Liss spoke of using nanotechnological devices to help restore ecosystems at a chemical level. A similar idea can be applied at a biological level.

The challenge—and it is huge—would be to develop insect-size or even microbe-size devices that could serve as selective, mobile, mechanical flyswatters or weed pullers. These could do what biological controls do, but would be unable to replicate and spread. Let's call devices of this sort "ecosystem protectors." They could keep aggressive imported species out, saving native species from extinction.

To a human being or an ordinary organism, an ecosystem protector would seem like just one more of the many billions of different kinds of bugs and microbes in the ecosystem—small things going about their own business, with no tendency to bite. They might be detectable, but only if you sorted through a lot of dirt and looked at it through a microscope, because they wouldn't be very common. They would have just one purpose: to notice when they bumped into a member of an imported species on the "not welcome here" list, and then either to eliminate it or to ensure, at least, that it couldn't reproduce.

Natural organisms are often very finicky about which species they attack. These ecosystem protectors could be equally finicky about which species they approach, and then, before attacking, could do a DNA analysis to be sure. It would be simplest (especially in the beginning while we're still learning) to limit each kind of defender to monitoring only one imported species.

Each unit of a particular kind of ecosystem-defender device would be identical, built with precision by a special-purpose molecular-manufacturing setup. Each would last for a certain time, then break down. Each kind can be tested in a terrarium, then a greenhouse, then a trial outdoors ecosystem, keeping an eye on their effects at each stage until one gains the confidence for larger-scale use. "Larger scale" could still be quite limited, if they aren't designed to travel very far. This built-in obsolescence limits both how long each device can operate and how far it can move: getting control of the structure of matter includes making nanomachines work where they're wanted and not work elsewhere.

The agricultural industry today manufactures and distributes many thousands of tons of poisonous chemicals to be sprayed on the land, typically in an attempt to eliminate one or a few species of insect. Ecosystem protectors could also be used to protect these agricultural monocultures, field by field, with far less harm to the environment than today's methods. They could likewise be used in the special ecosystems of intensive greenhouse agriculture.

Unlike chemicals sprayed into the environment, these ecosystem protectors would be precisely limited in time, space, and effect. They neither contaminate the groundwater nor poison bees and ladybugs. In order to weed out imported organisms and bring an ecosystem back to its natural balance, ecosystem protectors would not have to be very common—only common enough for a typical imported organism to encounter one once in a lifetime, before reproducing.

Even so, as the ecosystem protectors wear out and stop working, they would present a small-scale problem of solid-waste disposal. With the exercise of some clever design, all the machinery of ecosystem protectors might be made of reasonably durable yet biodegradable materials or (at worst) materials no more harmful than bits of grit and humus in the soil. So their remains would be like the shells of diatoms, or bits of lignin from wood, or like peculiar particles of clay or sand.

Alternatively, we might develop other mobile nanomachines to find and collect or break down their remains. This strategy starts to look like setting up a parallel ecosystem of mobile machines, a process that could be extended to supplement the natural cleansing processes of nature in many ways. Each step in this direction will require caution, but not paranoia: there need be no toxic chemicals here, no new creatures to spread and run wild. Missteps will have the great virtue of being reversible. If we decide that we don't like the effects of some particular variety of ecosystem protector or cleanup machine, we could simply stop manufacturing that kind. We could even retrieve those that had already been made and dispersed in the environment, since their exact number is known, along with which patch of ground each is patrolling.

If the making and monitoring of ecosystem protectors seems a lot of trouble to go to just to weed out nonnative species, consider this example of the environmental destruction such species can cause. Sometime before World War II, a South African species of fire ant was accidentally imported into the United States. Today, infested areas can have up to five hundred of these ants per square foot. The National Audubon Society—a strong opponent of irresponsible use of pesticides—had to resort to spraying its refuge islands near Corpus Christi when they found these ants destroying over half the hatchlings of the brown pelican, an endangered species.

In Texas, it's been shown that the new ants are killing off native ant species—reducing biodiversity. The USDA's Sanford Porter states that due to them, "Texas may be in the midst of a genuine biological revolution." The ants are heading west, and have established a beachhead in California. Without ecosystem protectors or something much like them, ecologies around the world will continue to be threatened by unnatural invasions. Our species opened the new invasion routes, and it's our responsibility to protect native species made newly vulnerable by them.

MENDING THE LAND

Today, most people are far from the land, tied up in turning the wheels of twentieth-century industry. In the years to come, those wheels will be replaced by molecular systems that do most of their turning by themselves. The pressure to destroy the land will be less. Time available to help heal the land will be greater. Surely, more energy will flow in this direction.

To mend ruined landscapes will require skill and effort. Ecosystem defenders can do flyswatting and weed-pulling jobs no humans ever could, but there will also be jobs of shaping, planting, and nurturing. The land has been torn by machines guided by hasty hands, almost overnight. It can gradually be restored by patient hands, whether bare, gloved, or guiding machines able to reshape a ravaged mountain without turning the soil.

The green wealth that can be brought by nanotechnology has raised high hopes among some environmentalists. Again writing in *Whole Earth Review*, Terence McKenna suggests it "would tend to promote . . . a sense of the unity and balance of nature and of our own human position within that dynamic and evolving balance." Perhaps people will learn to value nature more deeply when they can see it more clearly, with eyes unclouded by grief and guilt.

TEN

Nanomedicine

Our bodies are filled with intricate, active molecular structures. When those structures are damaged, health suffers. Modern medicine can affect the workings of the body in many ways, but from a molecular viewpoint it remains crude indeed. Molecular manufacturing can construct a range of medical instruments and devices with far greater abilities. The body is an enormously complex world of molecules. With nanotechnology to help, we can learn to repair it.

THE MOLECULAR BODY

To understand what nanotechnology can do for medicine, we need a picture of the body from a molecular perspective. The human body can be seen as a workyard, construction site, and battleground for molecular machines. It works remarkably well, using systems so complex that medical science still doesn't understand many of them. Failures, though, are all too common.

The Body as Workyard

Molecular machines do the daily work of the body. When we chew and swallow, muscles drive our motions. Muscle fibers contain bundles of molecular fibers that shorten by sliding past one another.

In the stomach and intestines, the molecular machines we call digestive enzymes break down the complex molecules in foods, forming smaller molecules for use as fuel or as building blocks. Molecular devices in the lining of the digestive tract carry useful molecules to the bloodstream.

Meanwhile, in the lungs, molecular storage devices called hemoglobin molecules pick up oxygen. Driven by molecular fibers, the heart pumps blood laden with fuel and oxygen to cells. In the muscles, fuel and oxygen drive contraction based on sliding molecular fibers. In the brain, they drive the molecular pumps that charge nerve cells for action. In the liver, they drive molecular machines that build and break down a whole host of molecules. And so the story continues through all the work of the body.

Yet each of these functions sometimes fails, whether through damage or inborn defect.

The Body as Construction Site

In growing, healing, and renewing tissue, the body is a construction site. Cells take building materials from the bloodstream. Molecular machinery programmed by the cell's genes uses these materials to build biological structures: to lay down bone and collagen, to build whole new cells, to renew skin, and to heal wounds.

With the exception of tooth fillings and other artificial implants, everything in the human body is constructed by molecular machines. These molecular machines build molecules, including more molecular machines. They clear away structures that are old or out of place, sometimes using machinery like digestive enzymes to take structures apart.

During tissue construction, whole cells move about, amoebalike: extending part of themselves forward, attaching, pulling their material along, and letting go of the former attachment site behind them. Individual cells contain a dynamic pattern of molecules made of components that can break down but can also be replaced. Some molecular machines in the cell specialize in digesting molecules that show signs of damage, allowing them to be replaced by fresh molecules made according to genetic instructions. Components inside cells form their complex patterns by self-assembly, that is, by sticking to the proper partners.

Failures in construction increase as we age. Teeth wear and crack and aren't replaced; hair follicles stop working; skin sags and wrinkles. The eye's shape becomes more rigid, ruining close vision. Younger bodies can knit together broken bones quickly, making them stronger than before, but osteoporosis can make older bones so fragile that they break under minor stress.

Sometimes construction is botched from the beginning due to a missing or defective genetic code. In hemophilia, bleeding fails to stop due to the lack of blood-clotting factor. Construction of muscle tissue is disrupted in 1 in 3,300 male births by muscular dystrophy, in which muscles are gradually replaced by scar tissue and fat; the molecule "dystrophin" is missing. Sickle-cell anemia results from abnormal hemoglobin molecules.

Paraplegics and quadriplegics know that some parts of the body don't heal well. The spinal cord is an extreme—and extremely serious—case, but scarring and improper regrowth of tissues result from many accidents. If tissues always regrew properly, injury would do no permanent physical damage.

THE BODY AS BATTLEFIELD

Assaults from outside the body turn it into a battlefield where the aggressors sometimes get the upper hand. From parasitic worms to protozoa to fungi to bacteria to viruses, organisms of many kinds have learned to live by entering the body and using their molecular machinery to build more of themselves from the body's building blocks. To meet this onslaught, the body musters the defenses of the immune system—an armada of its own molecular machines. Your body's own amoebalike white blood cells patrol the bloodstream and move out into tissues, threading their way between other cells, searching for invaders.

How can the immune system distinguish the hundreds of kinds of cells that should be in the body from the invading cells and viruses that shouldn't? This has been the central question of the complex science of immunology. The answer, as yet only partially understood, involves a complex interplay of molecules that recognize other molecules by sticking to them in a selective fashion. These include free-floating antibodies—which are a bit like bumbling guided missiles—and similar molecules that are bound to the surface of white blood cells and other cells of the immune system, enabling them to recognize foreign surfaces on contact.

This system makes life possible, defending our bodies from the fate of meat left at room temperature. Still, it lets us down in two basic ways.

First, the immune system does not respond to all invaders, or responds inadequately. Malaria, tuberculosis, herpes, and AIDS all have their strategies for evading destruction. Cancer is a special case in which the invaders are altered cells of the body itself, sometimes successfully masquerading as healthy cells and escaping detection.

Second, the immune system sometimes overresponds, attacking cells that should be left alone. Certain kinds of arthritis, as well as lupus and rheumatic fever, are caused by this mistake. Between attacking when it shouldn't and not attacking when it should, the immune system often fails, causing suffering and death.

MEDICINE TODAY

When the body's working, building, and battling go awry, we turn to medicine for diagnosis and treatment. Today's methods, though, have obvious shortcomings.

CRUDE METHODS

Diagnostic procedures vary widely, from asking a patient questions, through looking at X-ray shadows, through exploratory surgery and the microscopic and chemical analysis of materials from the body. Doctors can diagnose many ills, but others remain mysteries. Even a diagnosis does not imply understanding: doctors could diagnose infections before they knew about germs, and today can diagnose many syndromes with unknown causes. After years of experimentation and untold loss of life, they can even treat what they don't understand— a drug may help, though no one knows why.

Leaving aside such therapies as heating, massaging, irradiating, and so forth, the two main forms of treatment are surgery and drugs. From a molecular perspective, neither is sophisticated.

Surgery is a direct, manual approach to fixing the body, now practiced by highly trained specialists. Surgeons sew together torn tissues and skin to enable healing, cut out cancer, clear out clogged arteries, and even install pacemakers and replacement organs. It's direct, but it can be dangerous: anesthetics, infections, organ rejection, and missed cancer cells can all cause failure. Surgeons lack fine-scale control. The body works by means of molecular machines, most working inside cells. Surgeons can see neither molecules nor cells, and can repair neither.

Drug therapies affect the body at the molecular level. Some therapies—like insulin for diabetics—provide materials the body lacks. Most—like antibiotics for infections—introduce materials no human body produces. A drug consists of small molecules; in our simulated molecular world, many would fit in the palm of your hand. These molecules are dumped into the body (sometimes directed to a particular region by a needle or the like), where they mix and wander through blood and tissue. They typically bump into other molecules of all sorts in all places, but only stick to and affect molecules of certain kinds.

Antibiotics like penicillin are selective poisons. They stick to molecular machines in bacteria and jam them, thus fighting infection. Viruses are a harder case because they are simpler and have fewer vulnerable molecular machines. Worms, fungi, and protozoa are also difficult, because their molecular machines are more like those found in the human body, and hence harder to jam selectively. Cancer is the most difficult of all. Cancerous growths consist of human cells, and attempts to poison the cancer cells typically poison the rest of the patient as well.

Other drug molecules bind to molecules in the human body and modify their behavior. Some decrease the secretion of stomach acid, others stimulate the kidneys, many affect the molecular dynamics of the brain. Designing drug molecules to bind to specific targets is a growth industry today, and provides one of the many short-term payoffs that is spurring developments in molecular engineering.

LIMITED ABILITIES

Current medicine is limited both by its understanding and by its tools. In many ways, it is still more an art than a science. Mark Pearson of Du Pont points out, "In some areas, medicine has become much more scientific, and in others not much at all. We're still short of what I would consider a reasonable scientific level. Many people don't realize that we just don't know fundamentally how things work. It's like having an automobile, and hoping that by taking things apart, we'll understand something of how they operate. We know there's an engine in the front and we know it's under the hood, we have an idea that it's big and heavy, but we don't really see the rings that allow pistons to slide in the block. We don't even understand that controlled explosions are responsible for providing the energy that drives the machine."

Better tools could provide both better knowledge and better ways to apply that knowledge for healing. Today's surgery can rearrange blood vessels, but is far too coarse to rearrange or repair cells. Today's drug therapies can target some specific molecules, but only some, and only on the basis of type. Doctors today can't affect molecules in one cell while leaving identical molecules in a neighboring cell untouched because medicine today cannot apply surgical control to the molecular level.

NANOTECHNOLOGY IN MEDICINE

Developments in nanotechnology will result in improved medical sensors. As protein chemist Bill DeGrado notes, "Probably the first use you may see would be in diagnostics: being able to take a tiny amount of blood from somebody, just a pinprick, and diagnose for a hundred different things. Biological systems are already able to do that, and I think we should be able to design molecules or assemblies of molecules that mimic the biological system."

In the longer term, though, the story of nanotechnology in medicine will be the story of extending surgical control to the molecular level. The easiest applications will be aids to the immune system, which selectively attack invaders outside tissues. More difficult applications will require that medical nanomachines mimic white blood cells by entering tissues to interact with their cells. Further applications will involve the complexities of molecular-level surgery on individual cells.

As we look at how to solve various problems, you'll notice that some that look difficult today will become easy, while others that might seem easier turn out to be more difficult. The seeming difficulty of treating disorders is always changing: Once polio was frequent and incurable, today it is easily prevented. Syphilis once caused steady physical decline leading to insanity and death; now it is cured with a shot.

Athlete's foot has never been seen as a great scourge, yet it remains hard to cure. Likewise with the common cold. This pattern will continue: Deadly diseases may be easily dealt with, while minor ills remain incurable, or vice versa. As we will see, a mature nanotechnology-based medicine will be able to deal with almost any physical problem, but the order of difficulty may be surprising. Nature cares nothing for our sense of appropriateness. Horribleness and difficulty just aren't the same thing.

WORKING OUTSIDE TISSUES

One approach to nanomedicine would make use of microscopic mobile devices built using molecular-manufacturing equipment. These would resemble the ecosystem protectors and mobile cleanup machines discussed in the last chapter. Like them, they would either be biodegradable, self-collecting, or collected by something else once they were done working. Like them, they would be more difficult to develop than simple, fixed-location nanomachines, yet clearly feasible and useful. Development will start with the simpler applications, so let's begin by looking at what can be done without entering living tissues.

The skin is the body's largest organ, and its exposed position subjects it to a lot of abuse. This exposed position, though, also makes it easier to treat. Among the earlier applications of molecular manufacturing may be those popular, quasimedical products, cosmetics. A cream packed with nanomachines could do a better and more selective job of cleaning than any product can today. It could remove the right amount of dead skin, remove excess oils, add missing oils, apply the right amounts of natural moisturizing compounds, and even achieve the elusive goal of "deep pore cleaning" by actually reaching down into pores and cleaning them out. The cream could be a smart material with smooth-on, peel-off convenience.

The mouth, teeth, and gums are amazingly troublesome. Today, daily dental care is an endless cycle of brushing and flossing, of losing ground to tooth decay and gum disease as slowly as possible. A mouthwash full of smart nanomachines could do all that brushing and flossing do and more, and with far less effort—making it more likely to be used.

This mouthwash would identify and destroy pathogenic bacteria while allowing the harmless flora of the mouth to flourish in a healthy ecosystem. Further, the devices would identify particles of food, plaque, or tartar, and lift them from teeth to be rinsed away. Being suspended in liquid and able to swim about, devices would be able to reach surfaces beyond reach of toothbrush bristles or the fibers of floss. As short-lifetime medical nanodevices, they could be built to last only a few minutes in the body before falling apart into materials of the sort found in foods (such as fiber). With this sort of daily dental care from an early age, tooth decay and gum disease would likely never arise. If under way, they would be greatly lessened.

Going beyond this superficial treatment would involve moving among and modifying cells. Let's consider what can be done with this treatment inside the body, but outside the body's tissues. The bloodstream carries everything from nutrients to immune-system cells, with chemical signals and infectious organisms besides.

Here, it is useful to think in terms of medical nanomachines that resemble small submarines, like the ones in Figure 11. Each of these is large enough to carry a nanocomputer as powerful as a mid-1980s mainframe, along with a huge database (a billion bytes), a complete set of instruments for identifying biological surfaces, and tools for clobbering viruses, bacteria, and other invaders. Immune cells, as we've seen, travel through the bloodstream checking surfaces for foreignness and—when working properly—attacking and eliminating what should not be there. These *immune machines* would do both more and less. With their onboard sensors and computers, they will be able to react to the same molecular signals that the immune system does, but with greater discrimination. Before being sent into the body on their search-and-destroy mission, they could be programmed with a set of characteristics that lets them clearly distinguish their targets

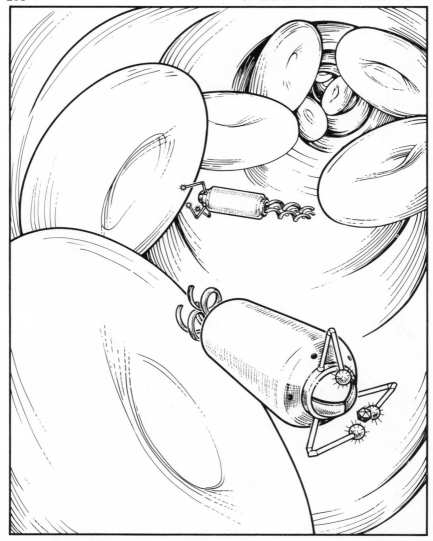

FIGURE 11: IMMUNE MACHINES

Medical nanodevices could augment the immune system by finding and disabling unwanted bacteria and viruses. The immune device in the foreground has found a virus; the other has touched a red blood cell. Adapted from Scientific American, January 1988.

from everything else. The body's immune system can respond only to invading organisms that had been encountered by that individual's body. Immune machines, however, could be programmed to respond to anything that had been encountered by world medicine.

Immune machines can be designed for use in the bloodstream or the digestive tract (the mouthwash described above used these abilities in hunting down harmful bacteria). They could float and circulate, as antibiotics do, while searching for intruders to neutralize. To escape being engulfed by white blood cells making their own patrols, immune machines could display standard molecules on their surface—molecules the body knows and trusts already—like a fellow police officer wearing a familiar uniform.

When an invader is identified, it can be punctured, letting its contents spill out and ending its effectiveness. If the contents were known to be hazardous by themselves, then the immune machine could hold on to it long enough to dismantle it more completely.

How will these devices know when it's time to depart? If the physician in charge is sure the task will be finished within, say, one day, the devices prescribed could be of a type designed to fall apart after twenty-four hours. If the treatment time needed is variable, the physician could monitor progress and stop action at the appropriate time by sending a specific molecule—aspirin perhaps, or something even safer—as a signal to stop work. The inactivated devices would then be cleared out along with other waste eliminated from the body.

WORKING WITHIN TISSUES

In most parts of the body, the finest blood vessels, capillaries, pass within a few cell diameters of every point. Certain white blood cells can leave these vessels to move among the neighboring cells. Immune machines and similar devices, being even smaller, could do likewise. In some tissues, this will be easy, in some harder, but with careful design and testing, essentially any point of the body should become accessible for healing repairs.

Merely fighting organisms in the bloodstream would be a major advance, cutting their numbers and inhibiting their spread. Roving medical nanomachines, though, will be able to hunt down invaders throughout the body and eliminate them entirely.

ELIMINATING INVADERS

Cancers are a prime example. The immune system recognizes and eliminates most potential cancers, but some get by. Physicians can recognize cancer cells by their appearance and by molecular markers, but they cannot always remove them all through surgery, and often cannot find a selective poison. Immune machines, however, will have no difficulty identifying cancer cells, and will ultimately be able to track them down and destroy them wherever they may be growing. Destroying every cancer cell will cure the cancer.

Bacteria, protozoa, worms, and other parasites have even more obvious molecular markers. Once identified, they could be destroyed, ridding the body of the disease they cause. Immune machines thus could deal with tuberculosis, strep throat, leprosy, malaria, amoebic dysentery, sleeping sickness, river blindness, hookworm, flukes, candida, valley fever, antibiotic-resistant bacteria, and even athlete's foot. All are caused by invading cells or larger organisms (such as worms). Health officials estimate that parasitic diseases, common in the Third World, affect more than one billion people. For many of these diseases, no satisfactory drug treatment exists. All can eventually be eliminated as threats to human health by a sufficiently advanced form of nanomedicine.

HERDING CELLS

Destroying invaders will be helpful, but injuries and structural problems pose other problems. Truly advanced medicine will be able to build up and restructure tissues. Here, medical nanodevices can stimulate and guide the body's own construction and repair mechanisms to restore healthy tissue.

What is healthy tissue? It consists of normal cells in normal patterns in a normal matrix all organized in a normal relationship to the surrounding tissues. Surgeons today (with their huge, crude tools) can fix some problems at the tissue level. A wound disrupts the healthy relationship between two different pieces of tissue, and surgical glues and sutures can partly remedy this problem by holding the tissues in a position that promotes healing. Likewise, coronary-artery bypass surgery brings about a more healthy overall configuration of tissues—one that provides working plumbing to supply blood to the heart muscle. Surgeons cut and stitch, but then they must rely on the tissue to heal its wounds as best it can.

Healing establishes healthy relationships on a finer scale. Cells must divide, grow, migrate, and fill gaps. They must reorganize to form properly connected networks of fine blood vessels. And cells must lay down materials to form the structural, intercellular matrix—collagen to provide the proper shape and toughness, or mineral grains to provide rigidity, as in bone. Often, they lay down unwanted scar tissue instead, blocking proper healing.

With enough knowledge of how these processes work (and nanoinstruments can help gather that knowledge) and with good enough software to guide the process—a more difficult challenge—medical nanomachines will be able to guide this healing process. The problem here is to guide the motion and behavior of a mob of active, living cells—a process that can be termed *cell herding*.

Cells respond to a host of signals from their environment: to chemicals in the surrounding fluids, to signal molecules on neighboring cells, and to mechanical forces applied to them. Cell-herding devices would use these signals to spur cell division where it is needed and to discourage it where it is not. They would nudge cells to encourage them to migrate in appropriate directions, or would simply pick them up, move them along, and deliver them where needed, encouraging them to nestle into a proper relationship with their neighbors. Finally, they would stimulate cells to surround themselves with the proper intercellular-matrix materials. Or—like the owner of a small dog who, on a cold day, wraps the beast in a wool jacket—they would directly build the proper surrounding structures for the cell in its new location.

In this way, cooperating teams of cell-herding devices could guide the healing or restructuring of tissues, ensuring that their cells form healthy patterns and a healthy matrix and that those tissues have a healthy relationship to their surroundings. Where necessary, cells could even be adjusted internally, as we will discuss later.

REBUILDING TISSUES

Again, skin provides easy examples and may be a natural place to start in practice. People often want hair where they have bare skin, and bare skin where they have hair. Cell-herding machines could move or destroy hair-follicle cells to eliminate an unwanted hair, or grow more of the needed cells and arrange them into a working follicle where a hair is desired. By adjusting the size of the follicle and the properties of some of the cells, hairs could be made coarser, or finer, or straighter, or curlier. All these changes would involve no pain, toxic chemicals, or stench. Cell-herding devices could move down into the living layers of skin, removing unwanted cells, stimulating the growth of new cells, narrowing unnaturally prominent blood vessels, insuring good circulation by guiding the growth of any needed normal blood vessels, and moving cells and fibers around so as to eliminate even deep wrinkles.

At the opposite end of the spectrum, cell herding will revolutionize treatment of life-threatening conditions. For example, the most common cause of heart disease is a reduced or interrupted supply of blood to the heart muscle. In pumping oxygenated blood to the rest of the body, the heart diverts a portion for its own use through the coronary arteries. When these blood vessels become constricted, we speak of coronary-artery disease. When they are blocked, causing heart muscle tissue to die, we speak of someone "having a coronary," another term for heart attack.

Devices working in the bloodstream could nibble away at atherosclerotic deposits, widening the affected blood vessels. Cell-herding devices could restore artery walls and artery linings to health, by ensuring that the right cells and supporting structures are in the right places. This would prevent most heart attacks.

But what if a heart attack has already destroyed muscle tissue, leaving the patient with a scarred, damaged, and poorly functioning heart? Once again, cell-herding devices could accomplish repairs, working their way into the scar tissue and removing it bit by bit, replacing it with fresh muscle fiber. If need be, this new fiber can be grown by applying a series of internal molecular stimuli to selected heart muscle cells to "remind" them of the instructions for growth that they used decades earlier during embryonic development.

Cell-herding capabilities should also be able to deal with the various forms of arthritis. Where this is due to attacks from the body's own immune system, the cells producing the damaging antibodies can be identified and eliminated. Then a cell-herding system would work inside the joint where it would remove diseased tissues, calcified spurs, and so forth, then rework patterns of cells and intercellular material to form a healthy, smoothly working, and pain-free joint. Clearly, learning to repair hearts and learning to repair joints will have some basic technologies in common, but much of the research and development will have to be devoted to specific tissues and specific circumstances. A similar process—but again, specially adapted to the circumstances at hand—could be used to strengthen and reshape bone, correcting osteoporosis.

In dentistry, this sort of process could be used to fill cavities, not with amalgam, but with natural dentin and enamel. Reversing the ravages of periodontal disease will someday be straightforward, with nanomedical devices to clean pockets, join tissues, and guide regrowth. Even missing teeth could be regrown, with enough control over cell behavior.

WORKING ON CELLS

Moving through tissues without leaving a trail of disruption will require devices able to manipulate and direct the motions of cells, and to repair them. Much remains to be learned—and will be easy to learn with nanoscale tools—but today's knowledge of cells is enough for a start on the problem of how to do surgery on cells.

Cell biology is a booming field, even today. Cells can be made to live and grow in laboratory cultures if they are placed in a liquid with suitable nutrients, oxygen, and the rest. Even with today's crude techniques, much has been learned about how cells respond to different chemicals, to different neighbors, and even to being poked and cut with needles. Conducting a rough sort of surgery on individual cells has been routine for many years in scientific laboratories.

Today, researchers can inject new DNA into cells using a tiny needle; small punctures in a cell membrane automatically reseal. But both these techniques use tools that on a cellular scale are large and clumsy—like doing surgery with an ax or a wrecking ball, instead of a scalpel. Nano-scale tools will enable medical procedures involving delicate surgery on individual cells.

Eliminating Viruses by Cell Surgery

Some viral diseases will respond to treatments that destroy viruses in the nose and throat, or in the bloodstream. The flu and common cold are examples. Many others would be greatly improved by this, but not eliminated. All viruses work by injecting their genes into a cell and taking over its molecular machinery, using it to produce more viruses. This is part of what makes viral illnesses so hard to treat—most of the action is performed by the body's own molecular machines, which can't be interfered with on a wholesale basis. When the immune system deals with a viral illness, it both attacks free virus particles before they enter cells, and attacks infected cells before they can churn out too many more virus particles.

Some viruses, though, insert their genes among the genes of the cell, and lay low. The cell can seem entirely normal to the immune system, for months or years, until the viral genes are triggered into action and begin the infective process anew. This pattern is responsible for the persistence of herpes infections, and for the slow, deadly progress of AIDS.

These viruses can be eliminated by molecular-level cellular surgery. The required devices could be small enough to fit entirely *within* the cell, if need be. Greg Fahy, who heads the Organ Cryopreservation Project at the American Red Cross's Jerome Holland Transplantation Laboratory, writes, "Calculations imply that molecular sensors, molecular computers, and molecular effectors can be combined into a device small enough to fit easily inside a single cell and powerful enough to repair molecular and structural defects (or to degrade foreign structures such as viruses and bacteria) as rapidly as they accumulate. . . . There is no reason such systems cannot be built and function as designed."

Equally well, a cell-surgery device located outside a cell could reach through the membrane with long probes. At the ends of the probes would be tools and sensors along with, perhaps, a small auxiliary computer. These would be able to reach through multiple membranes, unpackage and uncoil DNA, read it, repackage it, and recoil it, "proofreading" the DNA by comparing the sequences in one cell to the sequences of other cells.

On reading the genetic sequence spelling out the message of the AIDS virus, a molecular-surgery machine could be programmed to respond like an immune machine, destroying the cell. But it would seem to make more sense simply to cut out the AIDS virus genes themselves, and reconnect the ends as they were before infection. By doing this, and killing any viruses found in the cell, the procedure would restore the cell to health.

MOLECULAR REPAIRS

Cells are made of billions of molecules, each built by molecular machines. These molecules self-assemble to form larger structures, many in dynamic patterns, perpetually disintegrating and reforming. Cell-surgery devices will be able to make molecules of sorts that may be lacking, while destroying molecules that are damaged or present in excess. They will be able not only to remove viral genes, but to repair chemical and radiation-caused damage to the cell's own genes. Advanced cell-surgery devices would be able to repair cells almost regardless of their initial state of damage.

By activating and inactivating a cell's genes, they will be able to stimulate cell division and guide what types of cells are formed. This will be a great aid to cell herding and to healing tissues.

As surgeons today rely on the spontaneous, self-organizing ability of cells and tissues to join and heal the parts they manipulate, so cell-surgery devices will rely on the spontaneous self-organizing capabilities of molecules to join and "heal" the parts they put together. Healing of a surgical wound involves sweeping up dead cells, growing new cells, and a slow and genuinely painful process of tissue reorganization. In contrast, the joining of molecules is almost instantaneous and occurs on a scale far below that of the most sensitive pain receptor. "Healing" will not begin *after* the repair devices have done their work, as it does in conventional surgery: rather, when they complete their work, the tissue will have been healed.

HEALING BODY AND LIMB

The ability to herd cells and to perform molecular repairs and cell surgery will open new vistas for medicine. These abilities apply on a small scale, but their effects can be large scale.

CORRECTING CHEMISTRY

In many diseases, the body as a whole suffers from misregulation of the signaling molecules that travel through its fluids. Many are rare: Cushing's disease, Grave's disease, Paget's disease, Addison's disease, Conn's syndrome, Prader-Labhart-Willi syndrome. Others are common: Millions of older women suffer from osteoporosis, the weakening of bones that can accompany lowered estrogen levels.

Diabetes kills frequently enough to rank in the top ten causes of death in the United States; the number of individuals known to have it doubles every fifteen years. It is the leading cause of blindness in the United States, with other complications including kidney damage, cataracts, and cardiovascular damage. Today's molecular medicine tries to solve these troubles by supplying missing molecules: diabetics inject additional insulin. While helpful, this doesn't cure the disease or eliminate all symptoms. In an era of molecular surgery, physicians could choose instead to repair the defective organ, so it can regulate its own chemicals again, and to readjust the metabolic properties of other cells in the body to match. This would be a true healing, far better than today's partial fix.

Only now are researchers making progress on another frequent problem of metabolic regulation: obesity. Once this was thought to have one simple cause (consuming excess calories) and one main result (greater roundness than favored by today's aesthetics), but both assumptions proved wrong. Obesity is a serious medical problem, increasing the risk of diabetes mellitus, osteoarthritis, degenerative diseases of the heart, arteries, and kidneys, and shortening life expectancy. And the supposed cause, simple overeating, has been shown to be incorrect—something dieters had always suspected, as they watched thinner colleagues gorge and yet gain no weight.

The ability to lay in stores of fat was a great benefit to people once upon a time, when food supplies were irregular, nomadism and marauding bands made food storage difficult and risky, and starvation was a common cause of death. Our bodies are still adapted to that world, and regulate fat reserves accordingly. This is why dieting often has perverse effects. The body, when starved, responds by attempting to build up greater reserves of fat at its next opportunity. The main effect of exercise in weight reduction isn't to burn up calories, but to signal the body to adapt itself for efficient mobility.

Obesity therefore seems to be a matter of chemical signals within the body, signals to store fat for famine or to become lean for motion. Nanomedicine will be able to regulate these signals in the bloodstream, and to adjust how individual cells respond to them in the body. The latter would even make possible the elusive "spot reduction program" to reshape the distribution of body fat.

Here, as with many potential applications of nanotechnology, the problem may be solved by other means first. Some problems, though, will almost surely require nanomedicine.

New Organs and Limbs

So far we've seen how medical nanotechnology would be used in the simpler applications outside tissues—such as in the blood—then inside tissues, and finally inside cells. Consider how these abilities will fit together for victims of automobile and motorcycle accidents.

Nanomanufactured medical devices will be of dramatic value to those who have suffered massive trauma. Take the case of a patient with a crushed or severed spinal cord high in the back or in the neck. The latest research gives hope that when such patients are treated promptly after the injury, paralysis may be at least partially avoidable, sometimes. But those whose injuries weren't treated—including virtually all of today's patients—remain paralyzed. While research continues on a variety of techniques for attempting to aid a spontaneous healing process, prospects for reversing this sort of damage using conventional medicine remain bleak.

With the techniques discussed above, it will become possible to remove scar tissue and to guide cell growth so as to produce healthy arrangements of the cells on a microscopic scale. With the right molecular-scale poking and prodding of the cell nucleus, even nerve cells of the sorts found in the brain and spinal cord can be induced to divide. Where nerve cells have been destroyed, there need be no shortage of replacements. These technologies will eventually enable medicine to heal damaged spinal cords, reversing paralysis.

The ability to guide cell growth and division and to direct the organization of tissues will be sufficient to regrow entire organs and limbs, not merely to repair what has been damaged. This will enable medicine to restore physical health despite the most grievous injuries.

If this seems hard to believe, recall that medical advances have shocked the world before now. To those in the past, the idea of cutting people open with knives *painlessly* would have seemed miraculous, but surgical anesthesia is now routine. Likewise with bacterial infections and antibiotics, with the eradication of smallpox, and the vaccine for polio: each tamed a deadly terror, and each is now half-forgotten history. Our gut sense of what seems likely has little to do with what can and cannot be done by medical technology. It has more to do with our habitual fears, including the fear of vain hopes. Yet what amazes one generation seems obvious and even boring to the next. The first baby born after each breakthrough grows up wondering what all the excitement was about.

Besides, nano-scale medicine won't be a cure-all. Consider a fifty-year-old mentally retarded man, with a mind like a two-year-old's, or a woman with a brain tumor that has spread to the point that her personality has changed: How could they be "healed"? No healing of tissues could replace a missed lifetime of adult experience, nor can it replace lost information from a severely damaged brain. The best physicians could do would be to bring the patients to *some* physically healthy condition. One can wish for more, but sometimes it won't be possible.

FIRST AID

Throughout the centuries, medicine has been constrained to maintain functioning tissues, since once tissues stop functioning, they can't heal themselves. With molecular surgery to carry out the healing directly, medical priorities change drastically—function is no longer absolutely necessary. In fact, a physician able to use molecular surgery would prefer to operate on nonfunctioning, structurally stable tissue than on tissue that has been allowed to continue malfunctioning until its structure was lost.

Brain tumors are an example: They destroy the brain's structure, and with it the patient's skills, memories, and personality. Physicians in the future should be able to immediately interrupt this process, to stop the functioning of the brain to stabilize the patient for treatment.

Techniques available today can stop tissue function while preserving tissue structure. Greg Fahy, in his work on organ preservation at the American Red Cross, is developing a technique for vitrifying animal kidneys—making them into a low-temperature, crystal-free glass— with the goal of maintaining their structure such that, when brought back to room temperature, they can be transplanted. Some kidneys have been cooled to $-30°C$, warmed back up, and then functioned after transplantation.

A variety of other procedures can also stabilize tissues on a long-term basis. These procedures enable many cells—but not whole tissues—to survive and recover without help; advanced molecular repair and cell surgery will presumably tip the balance, enabling cells, tissues, and organs to recover and heal. When applied to stabilizing a whole patient, such a condition can be called *biostasis*. A patient in biostasis can be kept there indefinitely until the required medical help arrives. So in the future, the question "Can this patient be restored to health?" will be answered "Yes, if the patient's brain is intact, and with it the patient's mind."

Sandra Lee Adamson of the National Space Society has her eyes on distant goals. Some have proposed that travel to the stars would take generations, preventing anyone on Earth from ever making the trip. But she notes that biostasis will "give hope to some fearless adventurers who will risk suspension and subsequent reanimation so they can see the stars for themselves."

PLAGUE INSURANCE

Medical nanotechnologies promise to extend healthy life, but if history is any guide, they may also avert sudden massive death. The word *plague* is rarely heard today, except in relation to AIDS; it calls up visions of the Black Death of the Middle Ages, when one third of Europe died in 1346–50. A virulent influenza struck in 1918, half lost in the news of the First World War: how many of us realize that it killed at least 20 million? People often act as though plagues were gone for good, as if sanitation and antibiotics had vanquished them.

But as doctors are forever telling their patients, antibiotics kill bacteria, but are useless for viruses. The flu, the common cold, herpes, and AIDS—none has a really effective treatment, because all are caused by viruses. In some African countries, as much as 10 percent of the population is estimated to be infected with the AIDS-causing HIV virus. Without a cure soon, the steep rise in deaths from AIDS still lies in the future. AIDS stands as a grim reminder that the great plagues of history are not behind us.

THE THREAT

New diseases continue to appear today as they have throughout history. Today's population, far larger than that of any previous century, provides a huge, fertile territory for their spread.

Today's transportation system can spread viruses from continent to continent in a single day. When ships sailed or churned their way across the seas, an infected passenger was likely to show full-blown disease before arrival, permitting quarantine. But few diseases can be guaranteed to show themselves in the hours of a single aircraft flight.

So far as is known, every species of organism, from bacterium to whale, is afflicted with viruses. Animal viruses sometimes "jump the species gap" to infect other animals, or people. Most scientists believe that the ancestors of the AIDS virus could, until recently, infect only certain African monkeys. Then these viruses made the interspecies jump. A similar jump occurred in the 1960s when scientists in West Germany, working with cells from monkeys in Uganda, suddenly fell ill. Dozens were infected, and several died of a disease that caused both blood clots and bleeding, caused by what is now named the Marburg virus. What if the Marburg virus had spread with a sneeze, like influenza or the common cold?

We think of human plagues as a health problem, but when they hit our fellow species, we tend to see them from an environmental perspective. In the late 1980s, over half the harbor-seal population in large parts of the North Sea suddenly died, leading many at first to blame pollution. The cause, though, appears to be a distemper virus that made the jump from dogs. Biologists worry that the virus could infect seal species around the world, since distemper virus can spread by aerosols—that is, by coughing—and seals live in close physical contact. So far its mortality rate has been 60 to 70 percent.

What of AIDS itself: Could it change and give rise to a form able to spread, say, as colds do? Nobel Laureate Howard M. Temin has said, "I think that we can very confidently say that this can't happen." Nobel Laureate Joshua Lederberg, president of Rockefeller University in New York City, replied, "I don't share your confidence about what can and cannot happen." He points out that "there is no reason a great plague could not happen again. . . . We live in evolutionary competition with microbes—bacteria and viruses. There is no guarantee that we will be the survivors."

OUR INADEQUATE ABILITIES

Bacterial diseases are mostly controllable today. Sanitation limits the ways in which plague can spread. These measures are just good enough to lull us into imagining the problem is solved.

Viruses are common, viruses mutate; some spread through the air, and some are deadly. Plagues show that fast-spreading diseases can be deadly, and effective antiviral drugs are still rare.

The only really effective treatments for viral diseases are preventive, not curative. They work either by preventing exposure, or by exposing the body beforehand to dead or harmless or fragmentary forms of the virus, to prepare the immune system for future exposure. As the long struggle for an AIDS vaccine shows, one cannot count on modern medicine to identify a new virus and produce an effective vaccine within a single month or year or even a single decade. But influenza epidemics spread fast, and Marburg II or AIDS II or something entirely new and deadly may do the same.

DOING BETTER

The deaths from the next great plague could have begun in a village last week, or could begin next year, or a year before we learn to deal with new viral illnesses promptly and effectively. With luck, the plague will wait until a year after.

Immune machines could be set to kill a new virus as soon as it is identified. The instruments nanotechnology brings will make viral identification easy. Someday, the means will be in place to defend human life against viral catastrophe.

From eliminating viruses to repairing individual cells, improving our control of the molecular world will improve health care. Immune machines working in the bloodstream seem about as complex as some engineering projects human beings have already completed—projects like large satellites. Other medical nanotechnologies seem to be of a higher order of complexity.

ON SOLVING HARD PROBLEMS

Somewhere in the progression from relatively simple immune devices to molecular surgery, we've crossed the fuzzy line between systems that teams of clever biomedical engineers could design in a reasonable length of time and ones that might take decades or prove impossibly complex. Designing a nanomachine capable of entering a cell, reading its DNA, finding and removing a deadly viral DNA sequence, and then restoring the cell to normal would be a monumental job. Such tasks are advanced applications of nanotechnology, far beyond mere computers, manufacturing equipment, and half-witted "smart materials."

To succeed within a reasonable number of years, we may need to automate much of the engineering process, including software engineering. Today's best expert systems are nowhere near sophisticated enough. The software must be able to apply physical principles, engineering rules, and fast computation to generate and test new designs. Call it *automated engineering*.

Automated engineering will prove useful in advanced nanomedicine because of the sheer number of small problems to be solved. The human body contains hundreds of kinds of cells forming a huge number of tissues and organs. Taken as a whole (and ignoring the immune system), the body contains hundreds of thousands of different kinds of molecules. Performing complex molecular repairs on a damaged cell might require solving millions of separate, repetitive problems. The molecular machinery in cell-surgery devices will need to be controlled by complex software, and it would be best to be able to delegate the task of writing that software to an automated system. Until then, or until a lot of more conventional design work gets done, nanomedicine will have to focus on simpler problems.

AGING

Where does aging fit in the spectrum of difficulty? The deterioration that comes with aging is increasingly recognized as a form of disease, one that weakens the body and makes it susceptible to a host of other diseases. Aging, in this view, is as natural as smallpox and bubonic plague, and more surely fatal. Unlike bubonic plague, however, aging results from internal malfunctions in the molecular machinery of the body, and a medical condition with so many different symptoms could be complex.

Surprisingly, substantial progress is being made with present techniques, without even a rudimentary ability to perform cell surgery in a medical context. Some researchers believe that aging is primarily the result of a fairly small number of regulatory processes, and many of these have already been shown to be alterable. If so, aging may be tackled successfully before even simple cell repair is available. But the human aging process is not well enough understood to enable a confident projection of this; for example, the number of regulatory processes is not yet known. A thorough solution may well require advanced nanotechnology-based medicine, but a thorough solution seems possible. The result would not be immortality, just much longer, healthier lives for those who want them.

RESTORING SPECIES

A challenging problem related to medicine (and to biostasis) is that of species restoration. Today, researchers are carefully preserving samples from species now becoming extinct. In some cases, all they have are tissue samples. For other species, they've been able to save germ cells in the hope that they will be able to implant fertilized eggs into related species and thus bring the (nearly?) extinct species back.

Each cell typically contains the organism's complete genetic information, but what can be done with this? Many researchers today collect samples for preservation thinking only of the implantation scenario: one that they know has already been made to work. Other researchers are taking a broader view: the Center for Genetic Resources and Heritage at the University of Queensland is a leader in the effort. Daryl Edmondson, coordinator of the gene library, explains that the center is unique because it will "actively collect data. Most other libraries simply collate their own collections." Director John Mattick describes it as a "genetic Louvre" and points out that if genes from today's endangered species aren't preserved, "subsequent generations will see we had the technology to keep [DNA] software and will ask why we didn't do it." With this information and the sorts of molecular repair and cell-surgery capabilities we have discussed, lost species can someday be returned to active life again as habitats are restored.

One such center isn't enough: the Queensland center focuses on Australian species (naturally enough) and has limited funds. Besides, anything so precious as the genetic information of an endangered species should be stored in many separate locations for safety. We need to take out an insurance policy on Earth's genetic diversity with a broader network of genetic libraries, concentrating special attention on gathering biological samples from the fast-disappearing rain forests. Scientific study can wait: the urgency of the situation calls for a vacuum-cleaner approach. The Foresight Institute is promoting this effort through its BioArchive Project; interested readers can write to the address at the end of the Afterword.

ELEVEN

Limits and Downsides

The discussions of potential economic, medical, and environmental benefits may have given the false impression that nanotechnology will create a wondrous utopia in which all human problems are solved and we all live happily ever after. This is even more mistaken than the idea that new technologies always cause more problems than they solve. Many of the main constraints and difficulties faced by people are based not on technology or its lack, but instead by the very nature of the world we live in and the essence of our humanness.

Increasing affluence based on molecular manufacturing won't end economic problems any more than past increases in affluence have. Wilderness can still be destroyed; people can be oppressed; financial markets can be unstable; trade wars can be waged; inflation can soar; individuals, companies, and nations can go into debt; bureaucracy can stifle innovation; tax levels can become crippling; wars and terrorism can rage. None of these will automatically be stopped by advanced technology.

What is more, the potential benefits of new technologies aren't automatic. Nanotechnology *could* be used to restore the environment, to spread wealth, and to cure most illness. But will it? This depends on human action, working within the limits set by the real world.

This chapter first describes some of the limits to what nanotechnology can accomplish, and then some of the adverse side effects of its basically good applications. The next will discuss the problem of accidents, which seems manageable, and then the far greater problem of potential abuse of new capabilities.

SOME LIMITS OF NANOTECHNOLOGY

The world imposes limits on what we can do. Technology in general (and nanotechnology in particular) can provide padding for us as we throw ourselves against these hard, sharp limitations, and can sometimes help us slip past old limits through previously unknown gaps. Eventually, though, we will encounter new limits. In the end, solid constraints will limit human action no matter how much we juggle atoms and molecules, or the bits and bytes of information. Let's look at some of these, starting with the most abstract and long term—the most definite and hardest to avoid—and moving toward the more personal and near term.

INFORMATION LOSS

Many problems differ fundamentally from the material problems of limited matter and energy: they involve information. Some of the most precious stores of information in the world today are the genetic codes of the biosphere.

This information, different for virtually every individual organism, is the product of millions of events that we are incapable of modeling or recreating. When this information is lost, it is lost forever. When the atoms encoding this information are thoroughly scattered, there seems to be no way to retrieve it.

With any species, most genetic information is shared in common, found in all members of that species. But the variations in genetic code between individuals are important, both to the individuals themselves and to the health and prospects of the species as a whole. Consider the northern white rhino, whose numbers have dropped to an estimated thirty-two animals, or the California condor, of which only forty remain, all in captivity. Even if biologists succeed in reestablishing these species—eight condors were hatched in 1989— much of the diversity of their genetic information has been lost. Worse yet are extinctions of species for which no tissue samples were saved. The future may see some amazing recoveries: Dry skin and bones may yield a complete set of genes when sifted by molecular machinery, and even current techniques have been used to recover genes from an ancient leaf, almost 20 million years old. Our eyes and instruments cannot yet tell us how much information from the past remains, but we do know that genetic information is being lost every day, and once lost, it is irretrievable.

PHYSICAL LIMITS AND NONSENSE

People have often been wrong about physical limits, confusing the limits of their technology with the limits of the possible. As a result, learned men first dismissed the idea of heavier-than-air flight, and then dismissed the idea of flying to the Moon. Yet physical limits are real, and all technology—past, present, and future—will stay within those limits. There is even reason to suspect that some of those limits are where the learned now believe them to be.

Nanotechnology will make it possible to push closer to the real limits set by natural law, but it will not change those laws or the limits they set. It will not affect the law of gravity, the gravitational constant, the speed of light, the charge of the electron, the radius of the hydrogen atom, the value of Planck's constant, the effects of the uncertainty principle, the principle of least action, the mass of the proton, the laws of thermodynamics, or the boiling point of water. Nanotechnology won't make energy or matter from nothing.

It seems a good bet that no one will build a faster-than-light spacecraft, or an antigravity machine, or a cable twice as strong as diamond. There are limits. Science today may be wrong about some limits, but scientific knowledge is practically defined to be our best information about how the world works, so it isn't wise to bet against it.

There will be claims that nanotechnology will be able to do things that it can't, or that capabilities are around the corner when they aren't. Sometimes these will be innocent errors, sometimes they will be culpably stupid errors, and sometimes they will be what amounts to fraud. Among the problems that nanotechnology cannot solve is that of misguided claims, by people calling themselves "scientists," "engineers," or "businesspeople," that they have a big technical breakthrough worth a fortune. Every interesting new technology, particularly in its early days, is a chaotic mix of competent workers and charlatans. For every Thomas Edison inventing useful products such as light bulbs or the precursor of movie projectors, there were people promoting electric hairbrushes to cure baldness, and electric shoes, electric belts, electric hats—the list goes on—that authoritatively claimed cures for infertility, overweight, underweight, and all the ills and discomforts of mankind. Today, we laugh at the credulity of our forefathers who bought these gadgets; we shouldn't, unless we laugh at our own times as well.

POPULATION

Natural law imposes limits, but so does the nature of human beings. These will continue as long as people do.

Reproduction is a deeply ingrained instinct enforced by the march of time, which ruthlessly discards the genetic material of all who neglect it. Many would argue that the Earth is already overpopulated. While nanotechnology could enable the current population, and even a greatly increased one, to live more lightly on the Earth, there will still be limits to Earth's capacity.

The norms of human life are shaped by ancient patterns: high rates of infant or childhood mortality have been facts of life for millennia, and having many, many children has been a way to ensure that one or two will survive to work on the farm, and to care for you in your old age. Large families naturally become traditional. When modern medicine and reliable food supplies change those conditions—as they have, in cultural terms, virtually overnight—behavior does not shift as quickly. The result is the Third World population boom. In Western countries, where there has been time for behavior to adapt, a huge family is the exception.

It might seem that our problem is solved. Molecular manufacturing can make everyone wealthy, and wealthy populations today have stable or shrinking populations. The Earth can support more people with advanced technologies, and these will also open up the vast room and resources of the world beyond Earth. Would that this were true.

If 99 percent of the people in a population respond to wealth by reducing childbearing, the population will indeed stabilize or shrink, for a while. But populations are not uniform. What of the 1 percent, say, who are members of a minority with different values? If that minority has a growth rate of 5 percent per year, then in ninety-five years they will be the majority, and in one thousand years their population will have grown by a factor of 1,500,000,000,000,000,000,000, if resource limits or genocide haven't intervened. Note that the Hutterites of North America, a reasonably wealthy religious group viewing fertility control as a sin and high fertility as a blessing, have managed an average of ten children per woman. Given enough time, exponential growth of even the smallest population can consume all the resources in reach.

The right to reproduce is often regarded as basic, as illustrated by the outrage at reports of forced abortion in the People's Republic of China. The Hutterites and many others regard it as part of their freedom of religion. But what happens when parents have more children than they can support—does redistribution solve the problem? *If* reproduction is not forcibly suppressed, and *if* resources are forcibly *and repeatedly* redistributed so that each human being has a roughly equal share, then each person's share will steadily shrink. Even given the most optimistic assumptions regarding available resources, with a policy of resource redistribution and unlimited reproduction, the amount per person would eventually be insufficient to sustain life. This policy must be avoided, because if it is followed, it will kill everyone.

As soon as we grant that any entity is entitled to certain rights—whether that entity be a human child, an animal, or some future artificial intelligence—the question arises of who is responsible for providing resources to support it when it can't do so for itself. The above argument indicates that a policy of coercion by some central power to compel the entire population to support an exponentially exploding population of these individuals would lead directly to disaster. Ultimately, this responsibility must rest with the entities' initiator: the designer of the artificial intelligence, the owner of the pet, the parents of the child. No new technology can magically remove the limits imposed by natural law, and thereby lift the burden of human responsibility.

SOLUTIONS CAUSE PROBLEMS

Every time a technology solves a problem, it creates new problems. This doesn't mean that the change is neutral, or for the worse, of course. The Salk and Sabin vaccines for polio virtually destroyed the iron-lung industry, and the pocket calculator virtually destroyed the slide-rule industry, but these advances were worth the price of some economic adjustment.

Molecular manufacturing and nanotechnology will bring far greater changes, placing far greater strains on our ability to adapt. We shouldn't be surprised when basically beneficial applications make someone miserable. Our lives are largely centered around problems. If we can solve many of these problems, the centers of our lives will shift, creating fresh problems. This section sketches some of the issues of change and adaptation more to raise questions than to offer solutions.

CHANGE CAUSES PROBLEMS

Molecular manufacturing offers the possibility of drastic change, a change in the means of production more fundamental than the introduction of industry, or of agriculture. Our economic and social structures have evolved around assumptions that will no longer be valid.

How will we handle the changes in the way we work and live? Nanotechnology will have wide-ranging impact in many areas, including economic, industrial, and social patterns. What do historical patterns in similar circumstances tell us about the future?

Any powerful technology with broad applications revolutionizes lives, and nanotechnology will be no exception. Depending on one's point of view, this may sound exciting or it may sound disturbing, but it most certainly does not sound comfortable.

In comparison to many projections of the twenty-first century, though, nanotechnology may lead to *comparatively* comfortable change. The changes most often projected—for a future not including nanotechnology—have been ecological disaster, resource shortages, economic collapse, and a slide back into misery. The rise of nanotechnology will offer an alternative—green wealth—but that alternative will bring great changes from the patterns of recent decades.

Times of rapid technological change are disconcerting. For most of humanity's existence, people lived in a stable pattern. They learned to live as their parents had lived—by hunting and gathering, later by farming—and changes were small and gradual. A knowledge of the past was a reliable guide to the future.

Sudden changes, when they did occur, were apt to be ruinous: invasions or natural disasters. These sudden changes were fought or repaired or survived as best one could. Making major changes *by choice* was rare, and radical innovations were generally for the worse: the old ways at least ensured the ancestors' survival, the new might not. This made cultures conservative.

It is only natural that there be efforts to resist change, but before undertaking such an effort, it makes sense to examine the record of what works and what doesn't. The only examples of successful change fighters have been communities that have created and maintained barricades to isolate themselves from the outside world socially, culturally, and technologically. For the two centuries before 1854, Japan turned its back on the outside world, following a deliberate policy of seclusion. The leaders of Albania restricted contacts for many years; only recently have they started to open up.

Isolation attempts have worked better on a smaller scale, when participation is voluntary rather than decreed by government. Today, within the Hawaiian island chain, the tiny, privately owned island of Niihau, sixteen miles long and six miles wide, is deliberately kept as a preserve of the nineteenth-century Hawaiian lifestyle. Over two hundred full-blooded Hawaiians there speak the Hawaiian language and use no telephones, plumbing, television, and no electricity (except in the school). The Amish of Pennsylvania have no surrounding ocean to help maintain their isolation, but rely instead on tight social, religious, and technological rules aimed at keeping external technology and culture out, and themselves grouped in; those who leave the fold are excluded.

On a national scale, attempts to take only one part of the package—whether social or technological—haven't done well at all. For decades, the Soviet Union and the Eastern bloc nations welcomed Western technology but attempted tight restrictions on the passage of people, ideas, and goods. Yet illegal music, thoughts, literature, and other knowledge still crept in—as they do into the Islamic countries.

Fighting technological change in society at large has had little success, where that change gave some large group what it wanted. The most famous fighters of technological change—the Luddites—were unsuccessful. They smashed "automated" textile machinery that was replacing old hand looms during the early industrial revolution in England, but people wanted affordable clothing, and smashing equipment in one place just moved the business elsewhere. Change has sometimes been postponed, as when a later group, under the banner of "Captain Swing," smashed hundreds of threshing machines in a wide area of southern England in 1830. They succeeded in keeping the old, labor-intensive ways of harvesting for over a generation.

In previous centuries, when the world was less tightly connected by international trade, communications, and transportation, delays of years and even decades could be enforced through violence or legal maneuvers such as tariffs, trade barriers, regulations, or outright banning. Attempting to stop or postpone change is less successful today, when technology moves internationally almost as easily as people do—and human travel is so easy that 25 million people cross the Atlantic each year. Change fighters find that the problems they create mount with time. Products made using the old, high-cost techniques are uncompetitive. There is no way to bring back the "old jobs": they no longer make sense. But old habits die hard, and these same responses to the prospect of technological change continue today—ignoring it, denying it, and opposing it. Societies that have fought change, as Britain did, have fallen behind in a cloud of coal smoke.

Why did the Luddites respond violently? Perhaps their response can be attributed to three factors: First, the change in their lives was sudden and radical; second, it affected a large group of people at one time, in one area; and third, in a world unprepared for rapid technological change, there was no safety net to catch the unemployed. While local economies might have been able to absorb a trickle of hungry laid-off workers, they lacked the size and diversity needed to offer other employment options quickly to large numbers of unemployed.

In the twentieth century, however, societies have of necessity become somewhat better adapted to change. This has been a matter of necessity, because sluggish communities soon fall behind. In the ancient days of peasant stability, there was no need for institutions like *Consumer Reports* to study and rate new products, or regulators like the Environmental Protection Agency to watch over new hazards. We developed the needs, and we developed the institutions. These mechanisms represent important adaptations, not so much to the technologies of the twentieth century, but to the increasing *change* in technology during the twentieth century. There is great room for improvement, but they can perhaps provide a basis for adapting to the next century as well.

Even with the best of institutions to cushion shocks and discourage abuse, there will be problems. The very act of solving problems of production—of increasing wealth—will create problems of economic change.

CLEAN, DECENTRALIZED PRODUCTION CAUSES PROBLEMS

Over centuries, the trend has seemed to be toward centralization, beginning with the rise of factories and industrial towns. What drove these developments was the high cost of machinery and plant operations, the need to be near power sources, the impracticality of transportation among many small, dispersed sites, and the need for face-to-face communication.

Beginning with the first industrial revolution, factories employed large numbers of people in one place, leading to overcrowding and making local economies dependent on one industry and sometimes on a single company. Costly equipment necessitated central locations for textile production, rather than the cottage industries where a lone woman could earn a livelihood carding wool and creating thread on a spinning wheel (providing the origin of the term *spinster*). By the 1930s, the belief in the virtues of centralization and central planning—the supposed efficiencies and economies of scale—led to nationwide or continentwide experiments in centralization. But over the last decade, these large-scale experiments have been dismantled, from Britain's privatization of nationalized utilities to the beginning of a return to the market system in Eastern European countries.

Because the old limits on transportation, energy sources, and communication have fallen, business is now decentralizing. Between 1981 and 1986, the Forbes 500 companies cut their employees by 1.8 million. But during those same years, total civilian jobs went up by 9.2 million. Start-up companies created 14 million jobs; small companies created another 4.5 million. Telecommuting is booming, as are new businesses, independent professionals, and cottage industries.

We've also seen the resurgence of small, but highly diverse stores: gourmet-food shops, specialty ethnic shops, tea and coffee purveyors, organic and health-food stores, bakeries, yogurt shops, gourmet ice-cream stores, convenience stores offering twenty-four-hour access, shops selling packaged food plus snacks. These stores epitomize something fundamental: At some point, what we want is not a standard good at an ever-cheaper price, but special things customized to meet our own individual tastes or needs.

The trend for advanced technologies seems to be leading away from centralization. Will nanotechnology counter or accelerate this trend? By reducing the cost of equipment, by reducing the need for large numbers of people to work on one product, and bringing greater ability to produce the customized goods that people want, nanotechnology will probably continue the twentieth-century trend toward decentralization. The results, though, will be disruptive to existing businesses.

The computer industry perhaps provides a clue to what might happen as costs are lowered by nanotechnology. The computer-software industry is characterized by the garage-shop start-up. When your equipment is cheap—inexpensive PCs built around low-cost chips—and you can make a product by throwing in some ingenuity and human labor, it's possible to start a new industry on a shoestring.

In 1900, when cars were simple, there were many car manufacturers. By the 1980s, if you weren't an industrial giant like General Motors or Ford, Honda or Nissan, you had to be John De Lorean to even get a shot at acquiring the capital to play in the business. If molecular manufacturing can slash the capital costs for producing cars or other plant-intensive equipment, we will see the equivalent of garage-shop businesses springing up to offer new products, and hiring workers away from the industrial giants of today just as the personal computer has destroyed the dominance of the mainframe.

The American dream is to be an entrepreneur, and the technological trends of the twentieth century point in that direction. Nanotechnology probably continues it.

In one area, however, the late twentieth-century trend has been toward uniformity. The nations of Western Europe are in the process of uniting under one set of economic rules, and parts of Eastern Europe are anxious to join them. More and more supranational and transnational organizations knit the world together. The growth of trade has motivated economic integration.

Molecular manufacturing will work against this trend as well, permitting radical decentralization in economic terms. This will help groups that wish to step aside from the stream of change, enabling them to be more independent of the turbulent outside world, picking and choosing what technologies they use. But it will also help groups that wish to free themselves from the constraints of the international community. Economic sanctions will have little force against countries that need no imports or exports to maintain a high standard of living. And export restrictions will likewise do little to hamper a military buildup.

By weakening the ties of trade, molecular manufacturing threatens to weaken the glue that holds nations together. We need that glue, though, to deal with the arms-control issues raised by molecular manufacturing itself. This problem, caused by the potential for decentralization, may loom large in the coming years.

EVEN WEALTH AND LEISURE CAUSE PROBLEMS

Lester Milbrath, professor of sociology and political science, observes, "Nanotechnologies will create the problem of how to meaningfully and sustainably occupy the time of people who need not perform much work in order to have a sufficiency of life's goods. Our society has never faced this problem before, and it is not clear what social restructuring will be required to have a good society in those circumstances. We face much deep social learning."

The world has had little experience with what anthropologists call "abundance economies." The native American tribes of the Pacific Northwest were one of those rarities. Ruth Benedict, in her classic book *Patterns of Culture*, wrote, "Their civilization was built upon an ample supply of goods, inexhaustible, and obtained without excessive expenditure of labor." The Kwakiutls became famous for their "potlatches": contests in which they sought to shame their rivals by heaping more gifts upon them than they could ever return. The potlatches would often be a year in preparation, last for days, and occasionally involve destruction of entire buildings. It was certainly a colorful form of keeping up with the Joneses.

What will motivate us, once we have achieved an abundance economy? What will we regard as worthwhile goals to pursue? Increased knowledge, new art, improved philosophy, eliminating human and planetary ills? Will we find ourselves creating a better, wiser world, or sunk in boredom and jaded now that we have all and want nothing? If boredom gets out of hand, the lively spectacle of wealthy donors seeking to outdo each other to endow the arts, aid the poor, and do other good deeds for the sake of prestige would be welcome.

What will happen as life spans continue to lengthen and the time needed to make a living decreases? Even today, there are people who, when confronted with the prospect of a significantly longer life span, exclaim that they couldn't imagine what they would do with all that time. This response can be hard to understand, when it would take a thousand years to walk all the world's roads, more thousands of years to read all the world's books, and another ten thousand years to have a dinner conversation with each of the world's people—but tastes differ, and even a few decades of bad television might make anyone long for the peace of the grave.

CHANGING EMPLOYMENT CAUSES PROBLEMS

A major concern, and certainly the single area of greatest upheaval, is employment (which may become hard to distinguish from leisure). Once, people had little choice of employment. To keep a full belly, most had to work at the only job available: peasant farming. Eventually, people will have a complete choice of employment: they will be able to keep a full belly and a wealthy lifestyle while doing whatever they please. Today, we are about halfway between those extremes. In advanced economies, many different jobs are deemed useful enough that other people will offer an adequate income in exchange for the result. Some people can make a living doing something they enjoy—is this work, or leisure?

The impact of nanotechnology on patterns of employment will depend on when it arrives. Current demographics show a shrinking supply of young people entering the work force. Agriculture, the assembly line, and entry-level service jobs are experiencing a labor shortage, and no relief is in sight. If these trends continue, nanotechnology may show up in the midst of a shortage of labor. If it arrives late enough, it may compete with industries that are already nearing full automation; "job displacement" may mean replacing an industrial robot with a nanomachine.

Employment patterns have shifted radically in the past. One hundred and fifty years ago, the United States was an agricultural nation—69 percent of all people worked the land, and a growing percentage worked in industry doing things like building steam locomotives for Baldwin Locomotives Works or tanning leather for the giant Central Leather monopoly. By the early twentieth century, agriculture was waning in numbers but increasing in productivity; most people worked in industry, and the tiny information and service sector was beginning to grow. Today, the picture has reversed: 69 percent of employed Americans work on information or service jobs, only 28 percent work in industrial production, and 3 percent in agriculture. This tiny fraction feeds the other 97 percent of Americans, exports hugely to other countries, and receives subsidies and price-support payments to stop them from growing even more food. Manufacturing, even without nanotechnology, seems to be heading toward a similar condition.

With an ever-declining percentage of our population working in manufacturing, we have as everyday products things that were once available only to kings and the high nobility. Yet owning multiple suits of clothes, having personal portraits of ourselves and family members, having music upon our command, having a personal bedroom, and having a coach awaiting our need—these are now regarded as being among the bare necessities of life. It may be possible to adjust to even greater wealth with even less required labor, but the adjustment will surely cause problems. In a world in which nanotechnology reduces the need for workers in agricultural and manufacturing still further, the question will be asked, "What jobs are left for people to do once food, clothing, and shelter are very inexpensive?"

Again, the twentieth century provides some guidelines. As technology has reduced costs by efficiently producing many units of an identical item, people have begun to demand customization to meet individual needs or preferences. As a result, there are ever more jobs in producing custom goods. Today, semi-custom goods that try to help us meet our needs or express our taste abound: designer linens, ready-to-wear fashions, cosmetics, cars, trucks, recreational vehicles, furniture, carpeting, shoes, televisions, toys, sports equipment, washing machines, microwave ovens, food processors, bread bakers, pasta makers, home computers, telephones, answering machines—are all available in large and ever-changing variety.

Just as varied is the fabulous wealth and diversity of information produced in the twentieth century. Information products are a large factor in the economy: Americans buy 2.5 billion books, 6 billion magazines, and 20 billion newspapers each year. In recent years, new magazines have been invented and launched at the rate of one every business day of the year. A visit to a well-stocked magazine rack shows only a hint of the wealth of highly specialized publications, each one focused on a specialized interest or attitude: hotdog skiing, low-fat gourmet cooking, travel in Arizona, a magazine for people with a home office and a computer, and finely tuned magazines on health, leisure, psychology, science, politics, movie stars and rock stars, music, hunting, fishing, games, art, fashion, beauty, antiques, computers, cars, guns, wrestling.

Motion pictures, which started as a flock of independent production companies and then consolidated into the great studios of the 1930s, have since followed the decentralization and diversification trends of recent years. Now an expanding range of film entertainment comes via network TV, cable channels, private networks, videotapes, music videos. Independent producers are aided by the technology innovations of cable, direct-broadcast satellites, videotape technology, laser disks, videocameras.

The arts have burgeoned, with the general public as the new patron of the arts. Any artist or art form that could find and satisfy a market boomed in the twentieth century. Not just the traditional arts of actors, writers, musicians, and painters, but all forms of "domestic" artistry have grown to unprecedented levels: landscape and interior design, fashion design, cosmetics, hairstyling, architecture, bridal consulting.

Providing for these demands are some of the "service and information" jobs created in the late twentieth century. "Service" jobs include many ways of helping other people: from nursing to computer repairs to sales. In "information" jobs, projected to have the fastest percentage growth over the next decade, people find, evaluate, analyze, and create information. A magazine columnist or TV news producer obviously has an "information" job. But so do programmers, paralegals, lawyers, accountants, financial analysts, credit counselors, psychologists, librarians, managers, engineers, biologists, travel agents, and teachers.

"Increasingly," states *Forbes* magazine, "people are no longer laborers; they are educated professionals who carry their most important work tools in their heads. Dismissing them from their jobs, cutting them off from their places of employment may hurt them emotionally and financially. But it doesn't separate them from their vocation in the same way that pushing a farmer off his freshly seeded land does. For centuries workers were more dependent on a particular physical setting than they are now. Modern occupations generally give their practitioners more independence—and greater mobility— than did those of yesteryear."

These human skills that people carry with them will continue to be valued: managing complexity, providing creativity, customizing things for other people, helping people deal with problems, providing old services in new contexts, teaching, entertaining, and making decisions. A reasonable guess would be that many of the service and information industries of the twentieth century will continue to evolve and exist in a world with nanotechnology. What is harder to imagine would be what *new* industries will come into being once we have new capabilities and lower costs.

Along with the old economic law of supply and demand is another governing factor: price elasticity effects. Peoples' desire for something is "elastic": it expands or contracts when the cost of something valuable goes down or up. If the price of a flight to Europe is five hundred dollars, more people will take a European vacation than if the price is five thousand dollars. When you had to hire a highly trained mathematician to do equations, calculation was slow and expensive. People didn't do much of it unless they absolutely had to. Today, computers make calculation cheap and automatic. So now businesses do sophisticated financial modeling, chemists design protein molecules, students calculate orbital trajectories for spaceships, children play video games, moviemakers do ever more amazing special effects, and the cartoon—virtually extinct because of high labor costs—has returned to movie theaters, all because computers permit cheap calculation. Nanotechnology will offer new, affordable capabilities to these and other people. Today, it's as hard to predict what new industries will be invented as it would have been for the creators of the ENIAC computer to have predicted cheap, handheld game computers for children.

So rather than producing drastic unemployment, nanotechnology seems likely to continue the trend already seen today, away from jobs that can be automated and into jobs where the human perspective is vital. But the true possibilities are, as always in the modern world, beyond predicting.

Change Disrupts Plans

Major shifts in demographics always cause disruptions. Even when we know they are coming, we never prepare for them.

Our plans are based on expectations of what will happen. If things don't go as expected, we find that we have "malinvested." Houston real estate was valuable and looked to become even more so when times were good for the oil business there; when the fortunes of the oil business changed, Houston real estate was found to have been overbuilt, overpriced, and many millions of dollars were lost.

Lengthening life spans push people toward taking a longer-term perspective, but rapid rates of change force a shorter-term perspective in investments. Turbulence in technology and in governmental monetary policy have already shortened time horizons. Businesspeople once routinely built plants with a thirty-year useful life. Today, the rate of change is too fast, and uncertainty regarding inflation and potential changes in tax laws is too great for such investments to make sense. Faster change will shrink time horizons further.

Governments have taken on themselves the burden of looking a lifetime ahead, and the Social Security Administration is in for some rough times. When Otto von Bismarck, Germany's Iron Chancellor, came up with the notion of a guaranteed old-age pension, it was a cynically clever and low-cost way to gain popular goodwill. So few people lived to age sixty-five that the amounts paid out in pensions were a pittance. After watching the German experiment for a handful of years, other governments began following suit. None of them expected a world like ours where a baby girl born in the United States today has an *average* life expectancy of 78.4 years—double that of Bismarck's time—and even this estimate is based on the faulty assumption that her medical care will be no better than her great-grandmother's was.

At present, the Social Security Administration has two models: one they call "positive" and one they call "negative." In the "positive" model, people work like dogs until old age, retire, and promptly die—presumably before they've had a chance to collect substantial social-security or medical benefits. In the "negative" model, people retire early, develop illnesses that require medical intervention, and then live a long time making doctor visits and hospital stays during those years. Plans based on these models deserve to be disrupted. A better, more realistic scenario would have people living and able to support themselves for a long time, with illnesses that can be handled easily and inexpensively. Present social-security benefits are enough to provide a certain standard of living—food, housing, transportation, and so forth. In a future of great material wealth, these benefits will be easy to provide, and present projections of economic woe resulting from an aging population will seem quaint.

COPING WITH CHANGE

Back in the seventies, author Alvin Toffler brought out a book called *Future Shock*, describing how disturbing rapid change is for people. The book was a best-seller, but how much actual future shock has been seen in the past decade? Most people seem to have come through the last two decades pretty much all right, not in a state of shock at all. Rather than being shocked by technology, they are instead annoyed about pollution and traffic.

Does this mean Toffler was wrong in predicting future shock? It's true that technology has been advancing rapidly in many areas over the past twenty years. But consider the average person's home life: How much of this rapid technological advance has shown up there? A great deal, yet most of it is hidden, unlike the earlier part of the century, where obvious change was the norm. Electric lights and appliances, automobiles, telephones, airplanes, radio, and television affected almost everyone's private life. One person's life could span the time from horse-and-buggy travel to watching the Moon landings on television.

In contrast, the past twenty years have seen new technologies move more quietly into the home. The VCR and microwave oven don't seem nearly as revolutionary as earlier inventions. Telephone answering machines are useful but haven't caused major changes in lifestyles. Fax machines are handy, but they're much like having very fast mail, and as this is written, fax machines aren't yet in most homes. So it's not surprising that the average person has felt little future shock lately. New medicines taken as pills—which may be radically improved—look just like the earlier pills. The computerized bills that come in the mail aren't any more exciting to pay than the old human-prepared bills.

This situation is unlikely to last. How much longer can technology advance so rapidly in so many fields without major effects on our lifestyles? There's been a respite from future shock in the last three decades; people have had a chance to catch their breath. When nanotechnology arrives, will future shock arrive with it?

Some segments of society today are already getting practice in dealing with rapid technological advance. Those getting the most vigorous workout are in the computer field, where a machine two years old is regarded as obsolete, and software must be updated every few months to keep abreast of the new developments.

But has this terrific rate of progress been dizzying or overwhelming? Not for the consumer—on the contrary, computers have become easier to use. In the 1960s, the New Math that was introduced into American grade schools and junior high schools included extensive study of arithmetic using numbers written in something other than the familiar base 10. This was to prepare the "Adults of Tomorrow" for "the Computer Age" in which we would all be writing assembly-language computer programs in binary (base 2) code. But customers now purchase software rather than write it themselves— they need never deal with computer languages at all, much less a primitive assembly language. The rapid increase of computer speed has helped make computers easier to use.

This progression has occurred many times before: cars started off with external hand cranks, then advanced to starters you could yank from the comfort of the driver's seat; now starters perform invisibly when you turn the key in the ignition. This pattern will surely continue. First, some people will adapt to the technology, but in the long run, the technology will adapt to us. The most flexible and powerful the technology, the more easily it will adapt.

Seen from a distance, seemingly trivial patterns of adaptation form part of a larger process that has marked the last century: The Western world has begun to invent mechanisms to handle a world of persistent change. Our mechanisms are by no means perfect or painless, as any unemployed person can testify. Employment agencies and headhunters for job seekers; unemployment and severance packages to ease job transitions; on-the-job training, continuing education, retraining, specialized seminars to update professional skills, professional associations, networking, community resources centers, government training programs, and volunteer agencies are just a few of the inventions dealing with change and transition. Consumer information services, regulatory agencies, and environmental organizations are others. The most effective will endure. More options will continue to be invented.

TWELVE

Safety, Accidents, and Abuse

Some truisms: Almost any technology is subject to use, misuse, abuse, and accidents. The more powerful a technology is when properly used, the worse it is likely to be when abused. Any powerful technology in human hands can be the subject of accidents. Nanotechnology and molecular manufacturing will be no exception. Indeed, if molecular manufacturing replaces modern industry, and if its nanotechnological products replace most modern technologies, then most future accidents will have to involve nanotechnology.

Another truism: In a diverse, competitive world, any reasonably inexpensive technology with enormous commercial, medical, and military applications will almost surely be developed and used. It is hard to envision a scenario (short of the collapse of civilization) in which nanotechnology will not make its appearance; it seems inevitable. If so, then its problems, however tough, must be dealt with.

Like trucks, aircraft, biotechnology, rockets, computers, boots, and warm clothes, nanotechnology has the potential for both peaceful and aggressive uses. In peaceful uses (by definition), harm to people occurs either by accident or as an unintended consequence. In aggressive uses, harm is deliberate. In a peaceful context, the proper question to ask is *Can fallible people of goodwill, pursuing normal human purposes, use nanotechnology in a way that reduces risk and harm to others?* In an aggressive, military context, the proper question to ask is *Can we somehow keep the peace?* Our answer to the first will be a clear *yes*, and to the second, an apprehensive *maybe*.

Throughout this discussion, we assume that most people will be alert in matters concerning their personal safety, and that some will be alert in matters concerning world safety. During the 1970s, people awakening to the new large-scale, long-term problems of technology often felt isolated and powerless. They naturally felt that technology was out of their control, in the hands of shortsighted and irresponsible groups. Today, there are still battles to be fought, but the tide has turned. When a concern arises regarding a new, obvious technology, it is now much easier to get a hearing in the media, in the courts, and in the political arena. Improving these mechanisms for social vigilance and the political control of technology is an important challenge. Current mechanisms are imperfect, but they can still give a big push in the right directions.

Though we assume alertness, alertness can be a scarce resource. The total amount of concern and energy available for focusing on long-term problems is so limited that it must be used carefully, not squandered on problems that are trivial or illusory. Part of our aim in this chapter is to help sort out the issues raised by nanotechnology so that attention can be focused on problems that must be solved, but might not be.

The next few sections deal with accidents of conventional sorts, where safety benefits are obvious. Later sections discuss more novel problems, some tough enough that we have no good answers.

SAFETY IN ORDINARY ACTIVITIES

As countries have grown richer, their people have lived longer despite pollution and automobile accidents. Greater wealth means safer roads, safer cars, safer homes, and safer workplaces. Throughout history, new technologies have brought new risks, including risks of death, injury, and harm to the environment, but prudent people have only accepted new technologies when they offered an improved mix of risks and benefits. Despite occasional dramatic mistakes, the historical record says that people have succeeded in choosing technologies that reduce their personal risks. This must be so, or we wouldn't be living longer.

Molecular manufacturing and its products should continue this trend, not as an automatic consequence, but as a result of continued vigilance, of people exercising care in picking and choosing which technologies they allow into their daily lives. Nanotechnology will give better control of production and products, and better control usually means greater safety. Nanotechnology will increase wealth, and safety is a form of wealth that people value. Public debate, product testing, and safety regulations are standard parts of this process.

HOME SAFETY

In common home accidents, a dangerous product is wrongly applied, spilled, or consumed. Homes today are full of corrosive and toxic materials, for cleaning drains, dissolving stains, poisoning insects, and so forth. All too often, children drink them and die. With advanced technology, none of these tasks will require such harsh, crude chemicals. Cleaning could be performed by selective nanomachines instead of corrosive chemicals; insects could be controlled by devices like ecosystem protectors that know the difference between a cockroach and a person or a ladybug. There will doubtless be room for deadly accidents, but with care and hard work, it should be possible to ensure that nanotechnologies for the home are safer than what they replace, saving many lives.

It is, of course, possible to imagine safety nightmares: nanotechnology could be used to make products far more destructive than anything we've seen because it could be used to extend almost *any* ability further than we've seen. Such products presumably won't be commonplace: even today, nerve gas would make a potent pesticide, but it isn't sold for home use. Thinking realistically about hazards requires common sense.

Industrial Safety

We've already seen how post-breakthrough technologies can eliminate oil spills by eliminating oil consumption. A similar story could be told of almost any class of industrial accident today. But what about accidents—spills and the like—with the new technologies? Rather than trying to paint a picture of a future technology, of how it could fail and what the responses could be, it seems better to try a thought experiment. What could be done to deal with oil spills, if oil were still in use? This will show how nanotechnologies can be used to cope with accidents:

If there were a spill and oil on the shore, advanced nanomechanisms could do an excellent job of separating oil from sand, removing oil from rocks, and cleaning crude oil from feathers on birds and the feathery legs of barnacles. Oil contamination is a pollution problem, and nanotechnology will be a great aid in cleaning up pollution.

Buy why should the oil reach the shore? Economical production would make it easy to stockpile cleanup equipment near all the major shipping routes, along with fleets of helicopters to deliver it at the first distress call from a tanker. Oil cleanup equipment built with nanotechnology could surely do an excellent job of scooping oil from the water before it could reach the shore.

But why should the oil leave the tanker? Economical production of strong materials could make seamless hulls of fibrous materials far tougher than steel, with double, triple, or quadruple layers. Smart materials could even make punctures self-sealing. Hulls like this could be run into rocks at highway speeds without spilling oil.

But why should anyone be shipping crude oil across the sea? Even if oil were still being pumped (despite inexpensive solar energy and solar-derived fuels), efficient molecule-processing systems could refine it into pure, fluid fuels at the wellhead, and inexpensive tunneling machines could provide routes for deeply buried pipelines.

Any one of these advances would shrink or eliminate today's problem with oil spills, and all of them are feasible. This example suggests a general pattern. If nanotechnology can provide so many different ways to avoid or deal with an oil spill—one of the largest and most environmentally destructive accidents caused by today's industry—it can probably do likewise for industrial accidents in general.

The most direct approach is the most basic: the elimination of anything resembling today's bulk industrial plants and processes. The shift from messy drilling activities and huge tankers to small-scale distributed systems based on solar cells is characteristic of the style in which nanotechnology can be used. The chemical industry today typically relies on plants full of large, pressurized tanks of chemicals. Not surprisingly, these occasionally spill, explode, or burn. With nanotechnology, chemical plants will be unnecessary because molecules can be transformed in smaller numbers, as needed and where needed, with no need for high temperatures, high pressures, or big tanks. This will not only avoid polluting by-products, but reduce the risk of accidents.

MEDICAL SAFETY

Medicine can be safer too. Drugs often have side effects that can do permanent damage or kill. Nanomedicine will offer alternatives to flooding the body with a possibly toxic chemical. Often, one wants to affect just one target: just the stomach, or perhaps just the ulcer. An antibiotic or antiviral treatment should fight specific bacteria or viruses and not damage anything else. When medicine achieves the sophistication of immune machines and cell-surgery devices, this will become possible.

But what about medical accidents and side effects? Molecular manufacturing will make possible superior sensors to tell medical researchers of the effects of a new treatment, thereby improving testing. Better sensors will also help in monitoring any negative effects of a treatment on an individual patient. With care, only a few cells would be damaged and only small concentrations of toxic by-products would be produced before this was noticed and the treatment corrected.

The resources of nanotechnology-based medicine would then be available for dealing with the problem. With biostasis techniques available, even the worst medically induced illnesses could be put on hold while a treatment was developed. In short, serious medical mistakes could be made far rarer, and most mistakes could be corrected.

The conclusion that follows from these examples of oil spills, chemical plants, and the effects of medical treatments is straightforward. Today, our comparative poverty and our comparative technological incompetence press us in the direction of building and using relatively dangerous and destructive devices, systems, and techniques. With greater wealth and technological competence, we will have the option of accomplishing what we do today (and more) with less risk and less environmental destruction: in short, being able to do more, and do it better.

With better-controlled technologies, and with an ample measure of foresight and concern, we will even be able to do a better job of recovering from mistakes. It won't happen automatically, but with normal care we can arrange for our future accidents to be smaller and less frequent than those in our past.

EXTRAORDINARY ACCIDENTS

The previous section discussed ordinary accidents that would occur during the use of nanotechnology by generally responsible, yet fallible, human beings. Nanotechnology also raises the specter, however, of what have been termed "extraordinary accidents": accidents involving runaway self-replicating machines. One can imagine building a device about the size of a bacterium but tougher and more nearly omnivorous. Such runaways might blow like pollen and reproduce like bacteria, eating any of a wide range of organic materials: an ecological disaster of unprecedented magnitude—indeed, one that could destroy the biosphere as we know it. This may be worth worrying about, but can this happen *by accident?*

HOW TO PREPARE A BIG MISTAKE

The so-called Star Trek scenario (named after an episode of *Star Trek: The Next Generation* that featured runaway "nanites") is perhaps the most commonly imagined problem. In this scenario, someone first invests considerable engineering effort in designing and building devices almost exactly like the one just described: bacterial-sized, omnivorous, able to survive in a wide range of natural environments, able to build copies of themselves, and made with just a few built-in safeguards—perhaps a clock that shuts them off after a time, perhaps something else. Then, accidentally, the clock fails, or one of these dangerous replicators builds a copy with a defective clock, and away we go with an unprecedented ecological disaster.

This would be an extraordinary accident indeed. Note well, though, that this accident scenario starts with someone building a highly capable device that is *almost* disastrously dangerous, but held in check by a few safeguards. This would be like wiring your house with dynamite and relying on a safety-catch to protect the trigger: a subsequent explosion could be called an accident, but the problem isn't with the safety mechanism, it's with the dynamite installation.

Do we need to build nanotechnological dynamite? It's worth considering just how little practical incentive there is for anything even resembling the dangerous replicator just described. (Note that our topic here is accidents; deliberate acts of aggression are another matter.)

HOW TO AVOID IT

With our present technology, which is simpler to build—a car that runs on gasoline, or one that forages for fuels in the forest? A foraging car would be very hard to design, cost more to manufacture, and have more parts to break down. The situation is similar with nanotechnology.

Ralph Merkle of Xerox Palo Alto Research Center discussed this issue at the First Foresight Conference on Nanotechnology. He explains, "It's both uneconomical and more difficult to design a self-replicating system that manufactures every part it needs from naturally occurring compounds. Bacteria do this, but in the process they have to synthesize all twenty amino acids and many other compounds, using elaborate enzyme systems tailored specifically for the purpose. For bacteria facing a hostile world, the ability to adapt and respond to a changing environment is worth almost any cost, for lacking this ability they would be wiped out.

"But in a factory setting, where adequate supplies of all the needed parts are provided, the ability to synthesize parts from scratch is not only unneeded, it consumes extra time and energy, and produces excess waste. Even if we could design artificial self-replicating systems as flexible as existing natural ones, an inflexible and rigid system is better adapted to the controlled factory setting in which it will find itself than a more complex, more adaptable, less efficient design."

What is more, the Desert Rose Industries scenario showed how an expandable factory setup could operate with no self-replicating machines at all: molecular manufacturing doesn't require them. If they are used for some purpose, they will most likely resemble automobiles in their finicky requirements. A self-replicating molecular machine built for industrial purposes (and made as simple as possible) would float in a container of specially selected chemicals. As with the automobile, the best chemicals to use will probably be chemicals not commonly found in nature, and it would be easy to make that a design rule: *Never make a replicator that can use an abundant natural compound as fuel.*

If we follow this rule, the idea of a replicator "escaping" and replicating in the wild will be as absurd as the notion of an automobile going feral and refueling itself from tree sap. Whether for replicators or cars, to design a machine that could operate in the wild would not be a matter of a flick of the draftsman's pen, but an intense, sustained research-and-development effort focused on that objective. Crashes and explosions occur in machinery by accident, but complex new capabilities don't.

A simple psychological error frequently occurs when someone first hears about nanotechnology, and hears mention of "molecular machines," and "replicators," and "nanocomputers," and "nanomachines that operate in nature." The error is this: The person makes a single new mental pigeonhole for "nanotechnology," throws everything into it, and stirs. After some mental fermentation, the result is the mythical nanomachine that does everything: it's a replicator, it's a supercomputer, it's a Land Rover, it slices, it dices—and on reflection, this imaginary nanomachine sounds uncontrolled and dangerous. With enough effort, a do-it-all nanomachine could perhaps be built, but it sounds difficult and there's no good reason to try.

There *are* advantages to making systems of molecular machinery that can use inexpensive, abundant chemicals, and devices that can operate in nature, but these machines needn't be replicators. A facility like Desert Rose might be designed to use little but electric power from solar panels and molecules from the air, but a setup like this isn't going to slip away. Nanomachines built for cleaning up pollutants and other outdoor tasks could be manufactured in facilities run like Desert Rose and then spread or installed where they're needed.

Extraordinary accidents deserve attention, but with a little care they can be completely avoided. The incentive to build anything resembling a Star Trek–scenario replicator is negligible, even from a military perspective. Any effort toward building such a thing should be seen not as a use of nanotechnology, but as an abuse. Other abuses seem more likely, however, and are quite bad enough.

THE CHIEF DANGER: ABUSE

The chief danger of nanotechnology isn't accidents, but abuse. The safety benefits of nanotechnology, when used with normal care, will free some of our attention to grapple with this far more difficult problem. As Lester Milbrath observes, "Nanotechnologies have such great power that they could be used for evil or environmentally destructive purposes as easily as they could be used for good and environmentally nourishing purposes. This great danger will require a level of political control far beyond that which most nations know how to exercise. We have a prodigious social learning task that we must face."

Thus far, we've focused on how increased abilities can serve constructive ends. Not surprisingly, the potential consequences—with the huge exception of social and economic disruption—are overwhelmingly positive. Inherently clean, well-controlled, inexpensive, superior technologies, when applied with care, can yield far better results than inherently dirty, messy, costly, inferior technologies. This should come as no surprise, but it is only half of the story. The other half is the application of those same superior technologies to destructive ends.

Readers feeling that all this may be too good to be true can breathe a sigh of relief. This problem looks tough.

COOPERATIVE CONTROLS

Molecular manufacturing will lead to more powerful technologies, but our current, crude technology already has world-smashing potential. We have lived with that potential for decades now. In the coming years, we will need to strengthen institutions for maintaining peaceful security.

If most of the political power in the world, and with it most of the police and military power, sees that the course of self-interest lies in peace and stability, then solutions seem possible. (The prospect of an arms race in nanotechnology is terrifying and to be avoided at almost any cost. As of this writing, the end of the Cold War offers a better hope of avoiding this nightmare.) James C. Bennett, a high-tech entrepreneur and public-policy commentator affiliated with the Center for Constitutional Issues in Technology, explains the goal: "Advanced technologies, particularly as far-ranging a capability as nanotechnology, will create a strong demand for their regulation. The challenge will be to create sufficient controls to prevent the power-hungry from abusing the technologies, without either smothering development or creating an overbearing international regime."

In the coming decades, preventing major abuse of nanotechnology will take the form of regulation, arms control, and antiterrorist activities. In the field of arms control, nanotechnology should present strong motivation for international cooperation and for intimate mutual inspection in the form of joint-research programs.

The sheer productive capabilities of molecular manufacturing will make it possible to move from a working weapons prototype to mass production in a matter of days. In a more exotic vein, dangerous nanomachines could be developed, including programmable "germs" (replicating or nonreplicating) for germ warfare. Either development could bring war. With peace looking so profitable and an arms race looking so dangerous, arms control through cooperative development should look attractive. This does not make it easy, or likely.

Terrorism is not an immediate concern. We have lived with nuclear weapons and nerve gas for decades now, and nerve gas, at least, is not difficult to make. As of this writing, no city has been obliterated by terrorists using these means, and no terrorist has even made a credible threat of this sort. The citizens of Hiroshima and Nagasaki, like the Kurds in Iraq, fell victim to nuclear and chemical weapons wielded by governments, not small groups. So long as nanotechnology is technologically more challenging than the simple chemistry of nerve gas, nanoterrorism should not be a primary concern.

To keep dangerous nanotechnologies unavailable, however, will require regulation. If anyone were free to build anything using molecular manufacturing, then someday as the technology base improves and designs become available for more and more nanodevices, someone, somewhere—if only out of sheer spite—would figure out how to combine those nanodevices to make a dangerous replicator and turn it loose. There will almost surely be warning signs, however: In the natural course of events, causes attract protesters before stonethrowers, and produce letter bombs before car bombs. Abuse of nanotechnology is likely to be visible long before it is devastating, and this at least gives some time to try to respond.

REGULATORY TACTICS

Abuse of this sort can be delayed, perhaps for a long time, by proper regulation. The goal here isn't to make regulations so tight that people will have to violate them to get anything done. This would encourage holdouts, underground work, and disrespect for the law. Instead, the goal is to draw boundaries loosely enough to cause little difficulty for legitimate work, while making dangerous activities very difficult indeed. This is a delicate balance to strike: those fearful of risk naturally try to apply crude and oppressive regulations, and companies naturally try to loosen and avoid regulation entirely. Nonetheless, the problem must be solved, and this seems the best direction to explore.

In one approach, nanomachines could be divided into two classes: *experimental devices* and *approved products*. Approved products could be made widely available through special-purpose molecular manufacturing systems. Thus, once an experimental device had passed regulatory inspection, it could become inexpensive and abundant. In this way, popular demands for a product could be satisfied without anyone needing to break safety rules.

Approved products could include devices like (but superior to) the full range of modern consumer products, ranging from personal supercomputers with 3-D displays, through smart construction materials, to running shoes with truly amazing features. The main cost of such goods might be the royalty to the designer. In *Engines of Creation* (the first book to examine this topic), this strategy for producing and distributing approved products is called a "limited assembler system."

Note that both approved products and the limited assemblers that build them would lack the ability to make copies of themselves, to self-replicate. Ralph Merkle sees this ability as the one to keep an eye on: "Self-replicating systems can and should be appropriately regulated. There seems no need, however, to have any more than normal concerns for devices which cannot replicate. While we might, as with any device, need laws to ensure their appropriate use, they pose no extraordinary problems." For most products, normal medical, commercial, and environmental standards would apply; the regulatory bureaucracies are already in place.

There are great advantages to permitting nearly free experimentation in a new technology, allowing creative people to try ideas without seeking prior approval from a cumbersome committee. Surprisingly, this, too, seems compatible with safety.

In the world of nanotechnology, one cubic micron is a large space, with room enough for millions of components. For many purposes, a few cubic microns would amount to a large laboratory space. To a device on a micron scale, a centimeter is an enormous distance. Surrounding a micron-scale device with a centimeter-thick wall would be like surrounding a human being with a wall kilometers thick, and just as hard to penetrate. Further, a micron-scale device can be incinerated in an instant by something as small as a spark of static electricity. Based on observations like these, *Engines of Creation* outlined the idea of a *sealed assembler lab*, in which a researcher could build anything, even something deliberately designed to be dangerous, and yet be unable to get anything out of the microscopic sealed laboratory except for information.

With a good communications network, a researcher or product developer in Texas could equally easily perform experiments in a remote Maine laboratory run with the security and secrecy of a Swiss bank. A lab would have a responsibility to its customers to keep proprietary work confidential, and a responsibility to regulatory authorities to ensure that nothing but information leaves the laboratory. Researchers could then perform any small-scale experiments they wish. Only approved products, of course, would be built outside the sealed laboratories. While this may not be the best pattern of regulation possible, it does show one way in which freedom of experimentation could be combined with strict regulation of use. By providing a clear separation between legitimate and illegitimate activity, it would help with the difficult problem of identifying and preventing research aimed at damaging ends.

A sensible policy will have to balance the risk of private abuse of technology against the risk of government abuse of technology and regulation. Low-cost manufacturing can make surveillance equipment less expensive. Increased surveillance can reduce some risks in society, but the watchers themselves often aren't very well watched. Placing bounds on surveillance is a challenge for today's citizens as well as tomorrow's, and lessons learned in the past can be applied in the future.

In the long run, it seems wise to assume that someone, somewhere, somehow, will escape the bounds of regulation and arms control and apply molecular-manufacturing capabilities to making novel weapons. If by then we have had several decades of peaceful, responsible, creative development of nanotechnology (or perhaps a few years of help from smart machines), then we may have developed both ecosystem protectors and sophisticated immune machines for medicine. There is a good reason to think that distributed technologies of this sort could be adapted and extended to deal with the problem of protecting against novel nanoweaponry. Failure to do so could mean disaster. Nonetheless, building protective systems of this sort will be by far the greatest challenge of any we have discussed. The chief purpose of regulatory tactics like those we have described must be to buy time for those peaceful developments, to maximize the chances that this challenge can be met before time runs out.

(Any critic declaring this to be an optimistic book hereby stands charged with having failed to read and understand the above paragraph.)

GUIDE IT, OR STOP IT?

Potential accidents richly deserve the attention they will get, and we have confidence that this attention will suffice to make nanotechnology a force for improved human and environmental safety. Abuse is the greater danger, and harder to deal with. When considering a proposed policy, the first question should be, "How will this affect the long-term likelihood of serious abuse?"

Guiding Means Making Many Choices

Guiding a technology is a complex task. It means grappling with myriad decisions regarding which applications are beneficial and which harmful in such complex areas as medicine, the economy, and the environment. It means making such happy choices as which of several good approaches to apply in cleaning up toxic-waste dumps and reversing the greenhouse effect. It also means making more difficult choices in planning ecosystem restoration and environmental modification.

These problems will confront us with a range of choices better than we have today, yet choices that throw values into conflict. Which is a better use for a particular piece of land—the slow restoration of a wilderness ecosystem, or development as a recreational park? Either may be far better than pavement, strip mines, and dumps, but the choices will be controversial.

Likewise, in medicine, we will have a choice of developing many different ways to cure cancers, many different ways to cure heart disease, many different ways to cure AIDS. But the technologies that can be used to rebuild damaged heart muscle could be extended to rebuild muscle and connective tissue structures elsewhere in the body, without the harmful side effects of steroid drugs. The range of choices open to people will be enormous, and again will be cause for great debate.

When a new medical technology is discussed today, a frequent comment is, "This procedure raises ethical questions." This is often taken as a signal to delay its use, neglecting such ethical questions as "Is withholding this lifesaving treatment while we ponder akin to murder?" When a choice raises ethical questions or throws values into conflict, it is time to make an ethical decision or to step aside and let others choose for themselves. Deciding to avoid whatever raised the question is itself a decision—and often ethically indefensible. New technologies will face us with uncomfortable decisions, but so does life itself.

Setting up rules for nanotechnology development will be challenging: finding ways to maximize research freedom while preventing serious abuse *and making this stick worldwide* is a social challenge of the first rank. Beyond this are decisions regarding rules for its application, and the challenge of maximizing freedom of choice and action while preventing serious abuse, again *worldwide.*

To guide nanotechnology means grappling with a set of decisions that could ultimately remake much of the world—for the better if we are reasonably wise, or for the worse if we are too blundering and incautious. To avoid this responsibility (if we could) would be tempting, yet given the environmental and human stakes, it would, perhaps, be a wrong of historic proportions.

TRYING TO STOP MEANS LOSING CONTROL

The simplest imaginable approach to "guiding" nanotechnology would be to stop it. The easiest trip to plan is the trip that goes nowhere.

This would have a certain appeal, if it were possible. Because of its enormous potential for abuse, nanotechnology has the potential of doing great harm. If we believe that human beings and human institutions are too incompetent to deal with nanotechnology—that they are too likely to turn it to aggressive military use, or too likely to make it freely available to madmen—then the option of stopping the development of nanotechnology may seem attractive indeed. But the ethical question that must guide human actions is not "Would it be better to stop?," but "Would attempts to stop make things better?"

One option is to push forward, emphasizing the need for caution but also the potential for good applications. The promise of medical, economic, and environmental applications, joined with the threat posed by a new arms race, provides a powerful motive for international cooperation. With positive goals and an inclusive stance, international cooperation is a promising strategy; it could provide a basis for guiding the development and application of nanotechnology.

Another option would be to emphasize the downside, to focus

debate on potential abuses in support of a campaign to halt development. In following this strategy, an activist group would want to downplay the civilian applications of nanotechnology and emphasize its military applications. Horror stories of potential abuse (including abuses that regulation could easily prevent) would help to make the technology seem strange and dangerous.

This strategy might succeed in suppressing civilian research in many countries, though probably not all. Unfortunately, it would also guarantee funding for classified military research programs in laboratories around the world, even in the most morally honest countries, because of their then-inevitable fear of consequences if someone *else* developed nanotechnology first. In a hostile public atmosphere, research would be pushed into secret programs, and in secrecy the prospects for broad international cooperation would disappear. Attempts to stop nanotechnology for fear of a new, unstable arms race become self-fulfilling prophecies. Afterward, the advocates of this view could then say, "We warned you!" as the world slid toward a war they themselves had helped to prepare.

Attempting to stop technological development is a simple but dangerous idea. The greater its success, the greater the polarization it would cause between technology advocates and technology critics. A moderate success would push research out of the public universities and into corporate and military research labs. A greater success would push research out of the corporate laboratories and into heavily classified programs. A truly amazing success would end most of these, leaving the only remaining military programs in the hands of those states with thoroughly repressive governments or alien ideologies. This, presumably, is not how one would prefer nanotechnology to be developed.

The only genuine success would be a total success, and this would mean banning research not only in the United States, and Germany, and France, and the rest of Western Europe, and Japan, and the Soviet Union, and the People's Republic of China, and Taiwan, but in Korea, South Africa, Iran, Iraq, Israel, Brazil, Argentina, Vietnam, and the part of Colombia controlled by the Medellín Cartel. Later, as computers improve, as chemistry advances, as more and more proximal-probe microscopes are built by high school students, total success would require banning kids from tinkering in suburban garages in Pittsburgh.

Competitive pressures are pushing technology toward thorough control of matter, and we have seen that this goal can be reached by many different paths. Preventing one area of research would not prevent the advance, nor would stopping work in one country. When the United States delays drug development through regulation by the FDA, drug companies simply switch research overseas, or non-U.S. companies pull ahead. Orbital-launch capability and nuclear-weapons capability are other examples. Very seldom has one country given these abilities to another, yet at least eight nations are able to launch satellites to orbit independently, at least seven have detonated nuclear devices, and another two are suspected to be within reach of nuclear capability. India and Israel have built bombs and launched satellites, though neither is considered a great power or a leading force in world technology.

Where nanotechnology is concerned, many countries are capable of doing the required research, and more will be in the future. South Korea has both the needed educational levels and the ambition; visitors from the People's Republic of China ask about nanotechnology. A decision at the top directing the resources of a nation could get results almost anywhere. The United States is only gradually being shaken from its illusion that it rules the world of technology. This illusion is a poor basis for decisions and action.

RESPONSIBLE ACTION

For all practical purposes, nanotechnology seems inevitable. With work, it can be made beneficial, but only if we exercise ordinary care in avoiding accidents and extraordinary care in preventing abuse.

It's hard to get people to take future technologies seriously. Present-day problems dominate discussions, and ideas about future possibilities take effort to judge. Because of this inertia, broad international regulation of nanotechnology won't be possible until nanotechnology already exists, until people begin to see its results. And then, for regulation to be most effective, researchers and governments in many countries will need to cooperate and be on speaking terms with the technology's critics.

What, then, is the socially responsible course of action, the approach most likely to avoid serious abuse of nanotechnology and most likely to deliver some of its potential benefits? It is, we believe, to point out potential dangers and abuses and how they can be avoided, but also to emphasize the civilian applications in medicine, the environment, and the economy. It is these benefits that provide grounds for advocating open civilian development programs, and for international cooperation that can provide a basis for effective international guidance.

To guide nanotechnology will not be simple. We will be confronted with a range of choices greater than we have faced before in history. It is only by grappling with those choices that we will be able to affect them for the better.

THIRTEEN

Policy and Prospects

Although exploratory engineering research can show certain technological possibilities, gaining this knowledge can have a paradoxical effect on our *feeling* of knowledge, on our sense of how much we know about the future. It gives us more information, but it can reveal a range of possibilities so vast that we feel as if we know less than we did before.

The prospect of nanotechnology and molecular manufacturing has this paradoxical effect. It makes certain scenarios—such as a mid-twenty-first century world of poverty, or choking on pollution caused by massive accumulations of twentieth-century-style industry—seem very unlikely indeed. This is useful information in trying to understand our real situation and trying to make sensible plans for the future. And yet the range of new possibilities opened up is broader than we could have imagined before. On the negative side, one can imagine building engines of destruction capable of devastating the world as thoroughly as a nuclear war. On the positive side, one can imagine futures of stable peace with levels of health, wealth, and environmental quality beyond any historical precedent and beyond present expectations.

Within this spectrum of possibilities (and off to its sides) is a range of futures we can't even imagine. Our actions, day by day, are taking us into one of those futures. Not to some future of our present plans or dreams or nightmares, but to a real future, one that will grow from the intended and unintended consequences of our actions, one that we and our descendants will actually have to live in.

Scenarios are useful tools for thinking about the future. They don't represent predictions of what will happen, but instead they present pictures of worlds that one can imagine happening. By looking at these pictures and seeing how they fit together, we can try to get some idea of which events are more likely and which are less likely, and to get some idea of how the choices we make today may affect the shape of things to come.

SCENARIO 0: ORDINARY EXPECTATIONS (1990)

Nanotechnology will have little direct effect on the world until it is well developed, many years from now. The *expectation* of nanotechnology, however, is influencing how people think and act today. Yet even this expectation is still in the early stages of development and will likely have little effect on world affairs for years to come. In sketching scenarios, it seems sensible to begin with the standard worldview, at least for the next few years, and then to look at how nanotechnology and the expectation of nanotechnology might later begin interacting with large-scale developments.

As this is being written, old projections of East European, Middle Eastern, and world affairs have recently been upended, and expectations are fairly muddy. Still, one can identify the broad outlines of a conventional-wisdom view of expected events in the coming years and decades:

> Technology doesn't change much in the next five years, or indeed
> in the next fifty. Computer power continues to grow rapidly, but
> with few important effects. The great challenges of technology are

environmental: dealing with greenhouse gases and acid rain and the problems of toxic waste.

In parallel, more and more nations climb the ladder of technological capability to such thresholds as the ability to launch satellites, build nuclear weapons, and manufacture computer chips. With the worldwide flow of technical information and the worldwide emphasis on technological development, more and more second-rank countries follow close on the heels of the technological leaders.

Consumer electronics continues to improve, but this leads to a better-entertained population rather than a better-informed one. Exciting announcements like high-temperature superconductors and low-temperature fusion continue to appear, but after hearing cries of "Wolf!" and seeing only puppy dogs and fairy tales, most people discount news of purported breakthroughs.

Even in the thirty-to-fifty-year time frame, most newspaper stories and respected analysts assume there will be little technological change. Fifty-year projections of carbon-dioxide accumulation in the atmosphere assume that most energy will continue to come from fossil fuels. Thirty-year projections of economic crisis due to an aging population and a shrinking work force assume that economic productivity doesn't change greatly.

In terms of productivity and wealth, the United States continues to lose ground relative to the booming economies of Eastern Asia: to Japan, South Korea, Taiwan, and Singapore. In political terms, the Ordinary Expectations scenario is less clear, but expectations seem to run something like this: The breakup of the Eastern bloc and the collapse of communism as a "progressive" ideal lead to a freer and more democratic world. In Eastern Europe and perhaps in Central Asia, independent countries emerge, each with an industrial base and a population having substantial education in science and technology.

The relative decline of the United States economically and of the Soviet Union militarily loosen some of the ties that today bind the world's democracies to one another. The decreased threat of Soviet military power weakens alliances. As NATO loosens, and as the nations of Europe integrate their economic and political lives, gaps between the United States and Europe grow. As Soviet pressure on Japan weakens, the U.S.-Japanese military alliance weakens and trade frictions loom larger in comparison.

In this environment, protectionist pressures increase. An economic crash grows more likely. A shift from friendly relationships to peaceful hostility becomes an ominous possibility. The rise of multiple, nearly equal centers of economic and technological capability provides incentives for greater integration and cooperation, but also motives for great competition and secrecy.

In the long term, however, limited resources and the costs both of pollution and of pollution controls bring economic growth to a halt in an increasingly impoverished world. Population growth during this time has slowed, but creates great economic and environmental pressures. Resource conflicts escalate into war. The climate has changed irreversibly, the old forests are nearly gone, and extinction has swept a majority of species into nothingness.

Variations on the first five to ten years of the Ordinary Expectations scenario can provide a backdrop for scenarios covering the rise of nanotechnology in, perhaps, the next ten to twenty years:

SCENARIO 1: POLLYANNA TRIUMPHANT

We are living in a world like that of the Ordinary Expectations scenario where, after years of anticipation, primitive but fairly capable assemblers have recently been developed. For the first time, the media, the public, and policymakers take the prospect of nanotechnology seriously.

It looks very good to them. Technical work has shown that nanotechnology, once developed, can be used in a clean, controlled way, and that it can ultimately displace polluting industries while greatly increasing wealth per capita. The anticipated health benefits are enormous, and after years of a growing death toll from AIDS—only partially stemmed by advances in molecular medicine—the public has become very sensitive to the regular reports of human infection by exotic primate viruses from Africa. Concern about the stability of Earth's climate and ecosystems has grown as forests have shrunk and weather patterns have changed.

The prospect of breaking out of this cycle is appealing. It is clear that nanotechnology is no danger when in the hands of people of goodwill, and a relatively peaceful decade has allowed many people to forget the existence of other motives.

And so, with miraculously undivided popular support drawn from a grand coalition of environmentalists seeking to replace existing industry, industrialists seeking a more productive technology, health advocates seeking better health care, low-income groups seeking greater wealth, and so on and so forth, companies and governments plunge into nanotechnology with both feet and without reservation.

Development proceeds at a breakneck pace, and everyone who wants to participate in this great venture is welcome. Primitive assemblers are used to build better assemblers, which are used to build yet better assemblers, in laboratories and hobby shops around the world.

Products begin to pour forth. The economy is thrown into turmoil. Military equipment also begins to pour forth, and tensions begin to build. A military research group with more cleverness than sense builds a monster replicator, it eats everything, and we all die.

This scenario is absurd, at least in part because published warnings already exist. Since the 1960s, uncritical applause for new technologies has been limited to the now-defunct controlled presses of Eastern Europe (and similar places), and even there the resulting environmental disaster has become a matter for public debate, criticism, and correction.

In the expanding free world of today, the benefits, costs, and dangers of any great new technology will be thoroughly examined, expounded upon, and lied about from many different directions. We may or may not manage to make wise choices as a result. But one thing seems sure: Pollyanna will not triumph, because Pollyanna doesn't have the facts on her side.

SCENARIO 2: CHICKEN LITTLE RULES THE ROOST

Again, we are in the world of the Ordinary Expectations scenario, and primitive assemblers have recently been developed. Again, the prospect of nanotechnology is being taken seriously for the first time—but it is somehow portrayed as being just more of the same, but worse. Environmentalists view it not as an alternative to the polluting industries of the twentieth century, but as an extension of human power, and hence of the human ability to do harm. Horror stories of technology gone mad are spun to support this view.

Arms-control groups are justifiably alarmed by nanotechnology and emphasize its military applications. Groups seeking arms control via disarmament—and believing in unilateral strategies—work to prevent the development of nanotechnology everywhere they can, that is, everywhere within their political reach. To maximize their political leverage, they portray it as an almost purely military technology of immense and malign power.

Special-interest groups in industry see molecular manufacturing as a threat to their business and join the lobbying efforts to prevent it from happening. Unions, neglecting the prospect of greater wealth and leisure for their members, focus instead on possible disruptions in established jobs. They, too, oppose the development of the new technology. As a result, we hear not about how nanotechnology could be used in health care, environmental cleanup, and the manufacture of improved products, but about the insidious threat of tiny, uncontrollable military monster machines that will smash our industry.

After a few years of hearing this, public opinion in the industrial democracies is firmly "against the development of nanotechnology," but this is more a slogan than an enforceable policy. Laws are nevertheless passed to suppress it, and the focus of public debate returns to the old themes of poverty and disease and the newer themes of climatic change and environmental destruction. Solutions seem as distant as ever. No right-thinking person would have anything to do with nanotechnology, so only wrong-thinking people do.

But the initial debate hadn't become serious until assemblers were developed, and research had gone still further before the laws were passed. By then, nanotechnology was just around the corner.

Developing nanotechnology is primarily a matter of tools, just as was developing nuclear weapons. Decades earlier, nuclear-weapons capability had spread from one to two countries in forty-nine months, and to another three in the next fifteen years, despite the requirement for large quantities of exotic materials in each device. By the 1980s, there was already a huge international trade in chemical compounds, and many thousands of chemists who knew how to combine them to make new molecular objects, working not only in university research labs, in corporate research labs, and in civilian and military government research labs, but—as the black market in designer drugs shows—secretly, in criminal research labs.

Even in the 1980s, a scanning-tunneling microscope had been built as a high school science-fair project in the United States. There is nothing large-scale or exotic about synthetic chemistry or about precise positioning of molecules. And in our scenario, primitive assemblers have already been developed and techniques for constructing them published (as is standard practice) in the open scientific literature.

And so the attempts to suppress the development of nanotechnology succeed only in suppressing the *open* development of nanotechnology. But governments cannot be sure that other governments are not developing it in secret, and they have now heard so much about its military potential that this is impossible to ignore. Around the world, governments quietly set up secret research programs: some in democracies, others in the remaining authoritarian states.

There are even underground efforts. Once a primitive assembler or even an AFM-based molecular manipulator is in hand, the remaining challenges are chiefly those of design. In the 1980s, personal computers had become powerful enough to use for designing molecules. In the years since then, computer power has continued its exponential explosion. Peculiar elements of the technoculture join with—pick one: radical anarchists, radical reds, radical greens, or radical racists—in a project aimed at bringing down "the corrupt world order" of governments, of companies, of religions, of human beings, or of nonwhite/nonbrown people. With responsible groups out of the technology race, they see a real chance of finding the leverage needed to change the world.

And so years pass in comparative quiet, with occasional rumors of activity or exposure of a project. Then, from an unexpected direction beyond the reach of democratic control, destructive change breaks loose upon an unprepared world. The sky falls, and Chicken Little is vindicated.

With luck, we will find this scenario is also absurd. Public debate in the coming years will surely present a more balanced picture of the opportunities and dangers posed by the development of nanotechnology. Thoughtful people with conflicting views will become deeply involved. The impracticability of attempting to suppress technologies of this sort will likely become clear enough to give us a chance of keeping development in the open, in relatively responsible hands.

SCENARIO 3: INTERNATIONAL TECHNORIVALRY

A variant of the Ordinary Expectations scenario has played out for a number of years now. And after years of continuing turbulence, the net result is this: Japanese economic power has grown, with other East Asian economies beginning to close the gap. Their greater investment in long-range civilian R&D, with a focus since the late 1980s on engineering molecular systems, has enabled them to take the lead on the path to nanotechnology.

European economic integration and German unification, combined with the pressure of economic and technological competition from the United States and Japan, have turned Europe inward to some extent. Although cultural ties with the United States keep U.S.-European relations on a basically warm basis, hostility between Europe and Japan—already marked in the 1980s—has grown. Europe had long enjoyed great strength in chemistry and basic science, and in the 1980s had led the United States in organizing efforts on molecular electronics. This has placed them in a strong position with respect to nanotechnology, behind Japan but ahead of the United States.

The United States remains an enormously productive economy, but the cumulative effects of an educational system that neglects learning and corporations that emphasize quarterly results have made themselves felt. After decades of emphasizing the short term, people now find themselves living in the long term they had neglected. The reaction to U.S. relative economic decline has not been investment and renewal, but rhetoric and hostility directed toward "foreigners," particularly the Japanese.

It is thus an isolated and somewhat defensive Japan that builds the first molecular manipulator and recognizes its long-term potential. The technology is developed in a government-funded research laboratory with cooperation from major Japanese corporations. As the result of increasing tensions, foreign researchers—those still welcome in Japan—were not invited to participate in this particular effort.

A series of committee meetings formalizes a tacit decision made earlier in choosing researchers, and the specifics of this new development are treated as proprietary. Impressive results are announced, stirring pride in Japanese research, but the specifics of the methods involved are kept quiet.

This scarcely delays the diffusion of the basic technology. After the first demonstration, even the most myopic funding agencies support projects with the same goal. A European project had already been started in a French laboratory: it soon succeeds in building an assembler based on somewhat different principles. European researchers follow the Japanese precedent by keeping the details of their techniques as a loosely held secret, in the name of European competitiveness. The United States follows suit a year later in an effort funded by the Department of Defense.

Public life goes on much as before, dominated by the antics of entertainers and politicians, and by tales of the fate of the environment or the Social Security system in a fantasy-future of extrapolated twentieth-century technology. But more and more, in policy circles and in the media, there is serious discussion of nanotechnology and molecular manufacturing—what they mean and what to do about them.

In Japan, second-generation assemblers have begun to turn out small quantities of increasingly sophisticated molecular devices. These are prototypes of commercially useful products: sensors, molecular electronic devices, and scientific instruments; some are immediately useful even at a price of a hundred dollars per molecule. There are plans on the drawing boards for molecular assemblers that could make these devices at prices of less than one trillionth of a dollar. There are long-term plans (viewed with hope and anticipation) for full-fledged molecular manufacturing able to make almost anything at low cost from common materials.

This is exciting. It promises to at last free Japan from its decades-old dependence on foreign trade, foreign food, foreign raw materials, and foreign politics. By making spaceflight inexpensive and routine, it promises to open the universe to a people cooped up on a crowded archipelago. Investment soars.

Europe leads America but lags behind Japan and looks on Japanese progress with hostility. Europeans, too, share dreams of a powerful technology, and begin a race for the lead. The United States trails, but its huge resources and software expertise help it pick up speed as it joins the race. Other efforts also begin, and though they advance steadily, they cannot keep pace with the great power blocs.

On all sides, the obvious military potential of molecular manufacturing fires military interest, then research and development in both publicly announced and secret programs. Strategists play nanotechnology war games in their minds, in their journals, and on their computers. They come away shaken. The more they look, the more strategies they find that would enable a technologically superior power to make a safe, preemptive move—lethal or nonlethal—against all its opponents. Defenses seem possible in principle, but not in time.

Yet it becomes obvious that molecular manufacturing can provide defenses against lesser technologies. Even the great, mythical leak-proof missile shield looks practical when the defenders have vastly superior technology and a thousandfold cost advantage building military equipment.

No great power seems particularly hostile. By then, all have formally or informally been in a peaceful alliance for many years. Yet there are still memories of war, and the bonds of alliance and military cooperation are weakened by the lack of a common enemy and the growth of economic rivalry. And so squabbles over trade in obsolescing twentieth-century technologies poison cooperation in developing and managing the fresh technologies of the twenty-first century.

There are a thousand reasons to pursue military research and development in these technologies, and nationalistic economic competition helps keep that work secret on a nationalistic basis. Military planners must concern themselves not so much with intentions as with capabilities.

And so a technology developed in an atmosphere of commercial rivalry and secrecy matures in an atmosphere of military rivalry and secrecy. Advanced nanotechnologies arrive in the world not as advances in medicine, or in environmental restoration, or as a basis for new wealth, but as military systems developed in the midst of an accelerating multilateral arms race, with the quiet goal of preemptive use. Negotiations and development run neck and neck, and then . . .

SCENARIO 4: ENOUGH COHERENCE

Again our world is a variant of the Ordinary Expectations scenario, but the international environment is in a healthier condition. Despite trade friction, global economic integration has continued. Europe, the United States, and Japan all have a large stake in each other's well-being, and they recognize it. International military cooperation has continued, in part as a conscious counterweight to conflicts over trade. International cooperation in research has grown, spurred in part by the Japanese desire for closer international ties. The end of the Cold War has made secret military research programs less commonplace.

It is in this environment that primitive assemblers are developed, and it doesn't make a great difference who gets there first. As is standard in basic research, groups publish their results in the open literature and compete to impress their colleagues at home and abroad with the brilliance of their achievements.

The arrival of the first assemblers spurs serious debate on nanotechnology and its consequences, and that debate is reasonably open and balanced. It covers military, medical, and environmental consequences, with a major emphasis on how clean, efficient manufacturing can rise the level of wealth and spread it worldwide.

Military analysts consider the impact of molecular manufacturing and its potential products, and concerns are grave, so they undertake classified research programs. But—as usual—secrecy slows communication among researchers: those in the classified programs fall behind their more open colleagues, whose informal information-swapping runs far ahead of the published journals.

Some forces push toward rivalry; others push toward cooperation. A healthy pattern emerges: Those decision makers who take nanotechnology most seriously are precisely those who see the least reason for future international conflict among democratic nations. They no longer anticipate growing conflict over dwindling resources, inequalities of wealth, and global atmospheric pollution. They see what nanotechnology can do for these problems, without anyone taking anything from anyone else. And so, on all sides, those who take nanotechnology most seriously are those most inclined to look for cooperative solutions to the problems it poses. There are exceptions, but the tide of opinion is against them, and their ideas do not dominate policy.

The public debate on nanotechnology grows, and it ranges far and wide. Enthusiasts suggest many wondrous applications for nanotechnology. Some are soon dismissed as being impossible or just plain undesirable. Some are workable improvements on the horrid technologies of the twentieth century; these are developed and applied almost as soon as they become technically possible. The rest are harder to evaluate, but in the course of years of hard work and careful study some of these are developed and adopted, and others are rejected.

At first, some people proposed that nanotechnology be stopped, but they never proposed a credible way to do it. Realists observing the worldwide technological ferment look for other options to deal with the dangers.

The world's industrial democracies, taken together, hold the decisive lead. They have developed mechanisms for coordinating and controlling technologies with military potential by regulating technology transfer and trade. These mechanisms have been developed, exercised, and honed through decades of Cold War experience not only with nuclear and missile technologies, but with a host of high-technology products and devices. These mechanisms aren't perfect, but they are useful.

As concerns about international instability mount, the industrial democracies work to improve their teamwork: they reinforce the tradition of free trade and cooperation within the club, and strengthen regulations that block the flow of critical technologies to the world's remaining dictators.

As a result of these developments, nanotechnology matures in an atmosphere dominated more by economic cooperation than by military competition. The focus of policy is solidly on civilian applications, with due attention to potential military threats. Trust is reinforced by the automatic "mutual inspection" that is a natural part of cooperative research and development.

Hard decisions remain, and the shouting and the arguments grow louder throughout the world's media. But where the problem is clear, and survival or world well-being are at stake, necessary decisions are made and there is enough international coherence to implement them.

Years pass, and technologies mature. Health improves, wealth rises, and the biosphere begins to heal. Despite the turbulence and anguish of change—and despite voices saying, "It was better in the old days," at least for them, and despite real losses—many people of goodwill can look at the world, contemplate the whole, and affirm that this change is, on the whole, a change for the better.

PROSPECTS

Today's knowledge about molecules and matter is enough to give a partial picture of what molecular machines and molecular manufacturing will make possible. Even this partial picture shows possibilities that make old views of the twenty-first century thoroughly obsolete.

Science and technology are advancing toward molecular manufacturing along many fronts, in chemistry, physics, biology, and computer science. Motives for continuing range from the medical to the military to the scientific. Research in these directions is already worldwide, and just beginning to focus on the objective of nanotechnology.

Already, it is easy to describe how known devices and principles can be combined to build a primitive device able to guide molecular assembly. Actually doing it will not be so easy—laboratory research never is—but it will be done, and in not too many years.

The first, slow assemblers will lead to products that include better assemblers. Machines able to put molecules together to make molecular machines will lead to a spiral of falling costs and improving quality, ultimately yielding many results that people fervently want: a cleaner environment; an escape from poverty; health care that heals. These benefits will bring disturbing changes and unsettling choices, as new abilities always do. The pace of change may well accelerate, straining the institutions we have evolved to cope with turbulent times.

Molecular-manufacturing capabilities will lend themselves to abuse, and in particular, to the construction of weapons by those seeking power. To minimize the risk of such abuse, we need to develop broad-based international cooperation and regulation. Domestically, this focus seems the best way to avoid polarization between those concerned with solving old problems and those concerned with avoiding new ones. Internationally, it seems the best way to avoid a sickening slide into a new arms race.

As shown by the four scenarios just sketched, public opinion will shape public policy, helping to determine whether these technologies are used for good or for ill. The Afterword will look at today's state of opinion and at what can be done to push in a positive direction.

We cannot predict the future, and we cannot predict the consequences of our actions. Nonetheless, what we do will make a difference, and we can begin by trying to avoid every major blunder we can identify. Beyond this, we can try to understand our situation, weigh our basic values, and choose our actions with whatever wisdom we can muster. The choices we make in the coming years will shape a future that stretches beyond our imagining, a future full of danger, yet full of promise. It has always been so.

Taking Action

The human race is approaching the great historical transition to thorough, inexpensive control of the structure of matter, with all that implies for medicine, the environment, and our way of life. What happens before and during that transition will shape its direction, and with it the future.

Is this worth getting excited about? Look at some of the concerns that bring people together for action:

- Poverty
- Weapons systems
- Deforestation
- Toxic waste
- Social security
- Housing
- Global warming
- Endangered species
- Freedom
- Jobs
- Nuclear power
- Life extension
- Space development
- Acid rain
- AIDS, Alzheimers disease, heart disease, lung disease, cancer . . .

Each of these issues mobilizes great effort. Each will be utterly transformed by nanotechnology and its applications. For many of these issues, nanotechnology offers tools that can be used to achieve what people have been striving to accomplish. For many of these same issues, the abuse of nanotechnology could obliterate everything that has been achieved.

A good companion to the precept "Think globally, act locally" is "Think of the future, act in the present." If everyone were to abandon short-term problems and today's popular causes, the results would be disastrous. But there is no danger of that. The more likely danger is the opposite. The world is heading straight for a disruptive transition with everything at stake, yet 99.9 percent of human effort and attention is going into either short-term concerns or long-term strategies based on a fantasy future of lumbering twentieth-century technology.

What is to be done? For people more concerned with *feeling* good than with *doing* good, the answer is simple: Go for the warm feeling that comes from adding one more bit of support to an already-popular cause. The gratification is immediate, even if the contribution is small. For people more concerned with doing good—who can feel good only if they live up to their potential—the answer is less simple: To do the most good, find an important cause that is *not* already buoyed up by a cheering multitude, a project where one person's contribution almost automatically makes a big difference.

There is, today, an obvious choice for where to look. The potential benefits and drawbacks of nanotechnology generate a thousand areas for research, discussion, education, entrepreneuring, lobbying, development, regulation, and the rest—for preparation and for action. A person's contributions can range from career commitment to verbal support. Both can make a difference in where the world ends up.

OPINION MATTERS

What people do depends on what they believe. The path to a world prepared to handle nanotechnology begins with the recognition that nanotechnology is a real prospect.

What would be the response to a new idea as broad as nanotechnology, if it were true? Since it doesn't fall into any existing technical specialty, it wouldn't be anyone's job to provide an official, authoritative evaluation. Advanced molecular manufacturing can't be worked on in the lab today, so it wouldn't matter to scientists playing the standard careers-and-funding game. Still, some scientists and engineers would become interested, think about it, and lend support. *Science News*, covering the first major conference on the subject, would announce that "Sooner or later, the Age of Nanotechnology will arrive." This is, in fact, what has happened.

But what if the idea were false? Some curious scientists or engineer would soon point out a fatal error in the idea. Since the sweeping implications of nanotechnology make many people uncomfortable, a good counterargument would spread fast, and would soon be on the lips of everyone who would prefer to dismiss the whole thing.

No such counterargument has been heard. The most likely reason is that nanotechnology is a sound idea. Reactions have been changing from "That's ridiculous" to "That's obvious." The basic recognition of the issue is almost in place.

When nanotechnology emerges from the world of ideas to the world of physical reality, we will need to be prepared. But what does this require? To understand what needs to be done today, it is best to begin with the long term and then work back to the present.

WHERE WE WILL NEED TO BE

When the world is in the process of assimilating molecular manufacturing, years from now, it would be best if people were ready and if
the world situation favored peaceful, cooperative applications. Balanced international progress would be better than dominance by any
nation. Cooperative development would be better than technological
rivalry. A focus on civilian goals would be better than a focus on
military goals. A well-informed public supporting sound policies would
be better than a startled public supporting half-baked schemes.

All these goals will be best served if politicians aren't forced to act
like idiots—that is, if the state of public opinion permits them to
make the right decisions, and perhaps even makes bad decisions politically costly. The basic objective is straightforward: a world in which
as many people as possible have a basic understanding of what is
happening, a picture of how it can lead to a better future, and a
broad understanding of what to do (and not to do) to reach that future. The outlines of a positive scenario would then look something
like this:

Environmental groups and agencies have thought through the issues raised by nanotechnology, and know what applications they want
to promote and what abuses they want to prevent. Likewise, medical
associations, associations of retired persons, and the Social Security
Administration have thought through the issues raised by dramatically improved medical care and economic productivity, and are ready
with policy recommendations. Business groups have done likewise
with economic issues, and business watchdog groups are ready to
expose policies that merely serve special interests. Labor groups have
considered the impact of a deep, global economic restructuring on
the jobs and income of their members, and have proposals for cush-
ioning the shock without holding down productivity. Religious leaders have considered the moral dimensions of many applications, and
are ready with advice. Military analysts and arms-control analysts have
done the painstaking work of thinking through strategic scenarios,

and have developed an agreed-on core of policies for maintaining stability. International committees and agencies have made the new technologies a focus of discussion and planning, and backed by a healthy climate of opinion, they make international cooperation work.

Overall, supported by a framework of sensible public opinion and sensible politics, the complex process of adapting to change is working rather well. In field after field, group after group has put in the hard work needed to come up with policies that would advance their real interests without wrecking someone else's interests. This is possible more often than most would have expected, because molecular manufacturing makes possible so many positive-sum choices. There are still big battles, but there is also a large core of agreement.

In this time of transition, some people are actively involved in developing and guiding the technologies, but most people act as citizens, consumers, workers, friends, and family members. They shape what happens in the broader world by their votes, contributions, and purchases. They shape what happens in their families and communities by what they say, what they do, and especially by the educational investments they have made or supported. By their choices, they determine what nanotechnology means for daily life.

HOW WE CAN GET THERE

A world like this will require years of preparation. What can people do over the coming years to help this sort of world emerge, to improve the prospects for a peaceful and beneficial transition to new technologies? For the time being, the main task is to spread information.

People within existing organizations can nudge them toward evaluating nanotechnology and molecular manufacturing. A good start is to introduce others in the organization to the concepts, and talk through some of their implications. Follow-up activities will depend on the group, its resources, and its purposes.

For the time being, drafting of new regulations, lobbying of Congress, and the like all seem premature. Getting nanotechnology into the planning process, though, seems overdue. We invite existing organizations with concerns regarding medicine, the economy, the environment, and other issues of public policy to put nanotechnology on their agendas, and to join in debating and ultimately implementing sensible policies.

Some groups are doing relevant research work. Many could bias their choice of projects to favor goals in the direction of molecular systems engineering. For nanotechnology to be taken really seriously, some research group will have to build a reasonably capable molecular manipulator or a primitive assembler. This will require an interdisciplinary team, years of work, and a total cost unlikely to exceed one tenth that of a single flight of the U.S. Space Shuttle.

Other researchers can help by providing further theoretical studies of what advanced molecular manufacturing and nanotechnology will make possible. These studies can help groups know what to anticipate in their planning.

Some scientists and engineers will want to steer their careers into the field of nanotechnology. More students will want to study a combination of physics, chemistry, and engineering that will prepare them to contribute.

We encourage people of common sense and goodwill to become involved in developing nanotechnology. For those who have—or can gain—the necessary technical backgrounds, becoming involved with its development is an excellent way to influence how it is used. For better or for worse, technical experts in a field have a disproportionate influence over related policies.

During these years, there will be a growing need for grass-roots organizations aimed at public education and building a base for political action. Having a few thousand people ready to write five letters to Congress in some crucial year could make the difference between a world that works and a world destroyed by the long-term effects of a shortsighted bill.

What happens will depend on what people do, and what people do will depend on what they believe. The world is overwhelmingly shaped by the state of opinion: people's opinions about what will and won't happen, what will and won't work, what will and won't prove profitable or beneficial for themselves, for their families, for their businesses, for their communities, for the world. This state of opinion—as expressed in what people say to each other, and whether their actions conform to their words—shapes decisions day to day. During these years, it will matter greatly what people are saying to one another about the future, and how to make it work.

GETTING STARTED

With help from new technologies, we can renew the world—not make it perfect, not eliminate conflict, not achieve every imaginable dream, yet clear away many afflictions, both ancient and modern. With good preparation, we can perhaps even avoid creating too many new afflictions to take their place.

Who is responsible for trying to bring this about? Those who want to fight poverty, to earn their share of the benefits to come, to join in a great adventure, to meet people who care about the future, to save species, to heal the Earth, to heal the sick, to be at the cutting edge, to build international cooperation, to learn about technology, to fight dangers, to change the world—not necessarily all together, or all at once.

To help deal with the main problem today, lack of knowledge, you can encourage friends to read up on the subject. If you've liked this book, lend it.

The Foresight Institute publishes information and sponsors conferences on nanotechnology and its consequences. It provides a channel for news, technical information, and discussions of public policy, and it can help put you in contact with active people and organizations. To stay in touch with developments that will shape our future, please write or call:

> The Foresight Institute
> PO Box 61058
> Palo Alto, CA 94306
> 415-324-2490
> electronic mail: foresight@cup.portal.com

Further Reading

This lists further sources of nontechnical information on nanotechnology and related topics. (For more technical material, see the Technical Bibliography.)

FORESIGHT INSTITUTE

This nonprofit organization was founded to address the opportunities and challenges posed by nanotechnology and other powerful anticipated technologies. Materials available include the *Update* newsletter, the *Background* orientation series, occasional papers, and conference tapes. Students and others planning careers in nanotechnology-related research can request *Briefing #1: Studying Nanotechnology*. The Foresight Institute sponsors conferences on both technical and policy issues raised by nanotechnology. Readers concerned about endangered species should inquire about the BioArchive Project. The institute's address appears at the end of the Afterword.

ENGINES OF CREATION:
THE COMING ERA OF NANOTECHNOLOGY
by K. Eric Drexler

This book on nanotechnology (New York: Doubleday 1986) introduces the subject from a more abstract and long-term perspective. Topics covered include nanotechnology's relationship to scientific knowledge, the evolution of ideas, artificial intelligence, human life span, limits to growth, healing the environment, prevention of technological abuse, space development, and the need for new social technologies—such as hypertext publishing and fact forums—to help us deal with rapid technological change.

Available in Britain from Fourth Estate, and in Japan from Personal Media (under the title *Machines That Create: Nanotechnology*).

OTHER BOOKS AND ESSAYS

Atkins, P. W. *Molecules.* New York: Scientific American Library Series #21, 1987. An elegantly written and heavily illustrated introduction to the molecular world, showing many molecules in everyday use.

Bennett, James C. *Creating Competitive Space Trade: A Common Market for Space Enterprise.* Santa Monica, CA: Reason Foundation Policy Study No. 123, August 1990. Proposed a framework for international technology regulation that could be extended to nanotechnology.

Brand, Stewart. *The Media Lab: Inventing the Future at MIT.* New York: Viking, 1987. Vividly describes the lab's work on the personalized information technologies we'll be using tomorrow.

Burgess, Jeremy. *Microcosmos.* New York: Cambridge University Press, 1987. A collection of beautiful images of the microscale world.

Burnham, John C. *How Superstition Won and Science Lost.* New Brunswick, NJ: Rutgers, 1987. Tells the story of scientists' declining effort to reaching out to the public, and the resulting erosion of public understanding (which ultimately leads to flawed public policy).

Drexler, K. Eric. "Exploring Future Technologies," in *Doing Science: The Reality Club*, ed. John Brockman. New York: Prentice-Hall, 1991.

An essay describing the exploratory engineering approach to understanding future technological possibilities.

Drexler, K. Eric. "Technologies of Danger and Wisdom," in *Directions and Implications of Advanced Computing, Vol. 1*. Jonathan P. Jacky and Douglas Schuler, eds. Norwood, NJ: Ablex, 1989. This essay discusses how computer technologies could be used to strengthen social mechanisms for dealing with complex problems. The volume is based on the first major conference of the Computer Professionals for Social Responsibility.

Milbrath, Lester. *Envisioning a Sustainable Society*. Albany, NY: State University of New York Press, 1989. A broad work that includes a brief discussion of nanotechnology's potential effects.

Wildavsky, Aaron. *Searching for Safety*. New Brunswick, NJ: Transaction Publishers, 1988. This book documents how using new technologies can—and does—reduce old risks more rapidly than it creates new ones, and how either too little or too much caution can decrease safety.

ARTICLES AND MAGAZINES

Encyclopedia Britannica's Science and the Future Yearbook 1990. This annual includes an eighteen-page introduction to nanotechnology; offprints are available from the Foresight Institute (address appears in the Afterword).

"Computer Recreations." *Scientific American*, Jan. 1988. A column describing molecular mechanical computers.

"The Invisible Factory." *The Economist*, Dec. 9, 1989. A brief, clear, and technically accurate introduction to nanotechnology.

"Where the Next Fortunes Will be Made." *Fortune*, Dec. 5. 1988. Includes a discussion of the business consequences of nanotechnology.

Information and publications on biostasis and future medical capabilities are available from the Alcor Life Extension Foundation, 12327 Doherty Street, Riverside, CA 92503; telephone (714)-736-1703.

Science News is a weekly newsmagazine, accessible to the nontechnical reader. A good guide to (among other things) the latest developments on the path to nanotechnology.

Technical Bibliography

It's impossible to give a complete bibliography of nanotechnology-related publications here. The following books, papers, and articles will lead readers into some of the relevant literatures; a more complete list is available from the Foresight Institute (address in Afterword).

PAPERS AND ARTICLES

DeGrado, William F., Zelda R. Wasserman, and James D. Lear, "Protein Design, a Minimalist Approach." *Science* 243 (1989) 622–28. Describes successful work in protein design.

Drexler, K. Eric. "Molecular Engineering: An Approach to the Development of General Capabilities for Molecular Manipulation." *Proceedings of the National Academy of Sciences (USA)* 78 (1981) 5275–78. First description of flexible molecular manufacturing based on artificial molecular machinery.

Drexler, K. Eric, and John S. Foster. "Synthetic tips." *Nature* 343 (1990) 600. Proposes an approach to building a molecular manipulator.

Drexler, K. Eric. "Molecular Tip Arrays for AFM Imaging and Nanofabrication. *Journal of Vacuum Science and Technology B*. April 1991 (in press). An alternative approach to the goals described in "Synthetic Tips," aimed at sidestepping several technical problems and improving performance and flexibility.

Feynman, Richard. "There's Plenty of Room at the Bottom," a talk published in shorter form as "The Wonders that Await a Micro-microscope." *Saturday Review* 43 (April 2, 1960) 45–47; reproduced at greater length under its original title in *Miniaturization* ed. H. D. Gilbert New York: Reinhold, 1961. This visionary talk sketches top-down miniaturization to the microscale, and points clearly in the direction of nanotechnology.

Foster, J. S., J. E. Frommer, and P. C. Arnett. "Molecular Manipulation Using a Tunnelling Microscope," *Nature* 331 (1988) 324–26. Describes the first use of an STM for bonding molecules to a large object.

Huse, William D. et al., "Generation of a Large Combinatorial Library of the Immunoglobulin Repertoire in Phage Lambda." *Science* 246 (1989) 1275–81. Describes a method for generating protein molecules that bind other specific proteins by selecting from a large number of antibody fragments, without using mammalian cells.

Lehn, Jean-Marie. "Supramolecular Chemistry—Scope and Perspectives: Molecules, Supermolecules, and Molecular Devices." *Angewandte Chemie International Edition in English* 27 (1988) 89–112. Describes work in molecular recognition (Lehn's Nobel lecture).

Ponder, Jay W., and Frederic M. Richards. "Tertiary Templates for Proteins." *Journal of Molecular Biology* 193 (1987) 775–91. Describes computer-based methods for choosing amino-acid sequences compatible with a given folded structure.

BOOKS

Alberts, Bruce, et al. *Molecular Biology of the Cell, 2nd ed*. New York: Garland Publishing, 1989. Describes natural molecular machinery.

Burkert, Ulrich, and Norma L. Allinger. *Molecular Mechanics, ACS Monograph* 177 Washington, D.C.: American Chemical Society, 1982. The classic text on modeling molecules in mechanical terms, based on relationships between energy and molecular geometry.

Clark, Tim. *A Handbook of Computational Chemistry*, New York: Wiley-Interscience, 1985. Describes the use of computer-based classical and (especially) quantum mechanical models of molecules.

Crandall, B. C., and James Lewis, eds. *Proceedings of the First Foresight Conference on Nanotechnology* (working title). Cambridge, Mass.: MIT Press, scheduled for the late 1991.

Creighton, Thomas E. *Proteins.* New York: W. H. Freeman, 1984. An excellent introduction to proteins as physical objects.

Drexler, K. Eric. *Molecular Nanotechnology: Molecular Machines and Manufacturing* (working title, book in progress as of 1991). Presents the physical principles of molecular machinery, with an analysis of a basic set of devices.

Huberman, B. A. ed. *The Ecology of Computation.* Amsterdam: North-Holland/Elsevier Science Publishers, 1988. This collection includes three papers by Miller and Drexler presenting a market-based approach to organizing large-scale computation.

Maskill, Howard. *The Physical Basis of Organic Chemistry.* Oxford, Eng.: Oxford University Press, 1985. This is an unusual, useful textbook describing the chemistry of carbon-based molecules from the perspective of physical chemistry.

Rigby, Maurice, et al. *The Forces Between Molecules.* Oxford, Eng.: Clarendon Press, 1986. A good overview of its subject.

Finally, for those who object to the attempt to explain nanotechnology to the public at this early stage, see the book *How Superstition Won and Science Lost* by John C. Burnham (New Brunswick, NJ: Rutgers, 1987). It describes how scientists have abdicated their responsibility in this area, and some of the consequences.

TAPES

Audio and videotapes are available from the First Foresight Conference on Nanotechnology held in Palo Alto, California, in October 1989. Contact the Foresight Institute, P.O. Box 61058, Palo Alto, CA, 94306; telephone (415) 324-2490.

ERATO BROCHURE

Descriptions of current research projects in the Exploratory Research for Advanced Technology program (ERATO) are available from: Research Development Corporation of Japan, 5-2, Nagata-cho 2-chome, Chiyoda-ku, Tokyo 100, Japan; fax 03-581-1486.

FOR STUDENTS

See also the nontechnical Further Reading section, and especially the Foresight Institute publications. Major advances in the enabling sciences are often published in the journals *Science* and *Nature*, both of which are worth browsing every week.

Glossary

Some terms used in discussing nanotechnology and other anticipated technologies:

Assembler: A general-purpose device for molecular manufacturing capable of guiding chemical reactions by positioning molecules.

Atom: The smallest unit of a chemical element, about a third of a nanometer in diameter. Atoms make up molecules and solid objects.

Atomic force microscope (AFM): An instrument able to image surfaces to molecular accuracy by mechanically probing their surface contours. A kind of proximal probe.

Automated engineering: Engineering design done by a computer system, generating detailed designs from broad specifications with little or no human help.

Automated manufacturing: As used here, nanotechnology-based manufacturing requiring little human labor.

Bacteria: Single-celled microorganisms, about one micrometer (one thousand nanometers) across.

Bulk technology: Technology in which atoms and molecules are manipulated in bulk, rather than individually.

Cell: A small structural unit, surrounded by a membrane, making up living things.

Cell pharmacology: Delivery of drugs by medical nanomachines to exact locations in the body.

Cell surgery: Modifying cellular structures using medical nanomachines.

Disassembler: An instrument able to take apart structures a few atoms at a time, recording structural information at each step.

DNA: A molecule encoding genetic information, found in the cell's nucleus.

Ecosystem protector: A nanomachine for mechanically removing selected imported species from an ecosystem to protect native species.

Enabling science and technologies: Areas of research relevant to a particular goal, such as nanotechnology.

Enzymes: Molecular machines found in nature, made of protein, which can catalyze (speed up) chemical reactions.

Exploratory engineering: Design and analysis of systems that are theoretically possible but cannot be built yet, owing to limitations in available tools.

Gray goo: See **Star Trek scenario**.

Immune machines: Medical nanomachines designed for internal use, especially in the bloodstream and digestive tract, able to identify and disable intruders such as bacteria and viruses.

Limited assembler: Assembler capable of making only certain products; faster, more efficient, and less liable to abuse than a general-purpose assembler.

Molecular electronics: Any system with atomically precise electronic devices of nanometer dimensions, especially if made of discrete molecular parts rather than the continuous materials found in today's semiconductor devices.

Molecular machine: Any machine with atomically precise parts of nanometer dimensions; can be used to describe molecular devices found in nature.

Molecular manipulator: A device combining a proximal-probe mechanism for atomically precise positioning with a molecule binding site on the tip; can serve as the basis for building complex structures by positional synthesis.

Molecular manufacturing: Manufacturing using molecular machinery, giving molecule-by-molecule control of products and by-products via positional chemical synthesis.

Molecular medicine: A variety of pharmaceutical techniques and therapies in use today.

Molecular nanotechnology: Thorough, inexpensive control of the structure of matter based on molecule-by-molecule control of products and by-products; the products and processes of molecular manufacturing, including molecular machinery.

Molecular recognition: A chemical term referring to processes in which molecules adhere in a highly specific way, forming a large structure; an enabling technology for nanotechnology.

Molecular surgery or molecular repair: Analysis and physical correction of molecular structures in the body using medical nanomachines.

Molecular systems engineering: Design, analysis, and construction of systems of molecular parts working together to carry out a useful purpose.

Molecule: Group of atoms held together by chemical bonds; the typical unit manipulated by nanotechnology.

Nano-: A prefix meaning one billionth (1/1,000,000,000).

Nanocomputer: A computer with parts built on a molecular scale.

Nanoelectronics: Electronics on a nanometer scale, whether made by current techniques or nanotechnology; includes both molecular electronics and nanoscale devices resembling today's semiconductor devices.

Nanomachine: An artificial molecular machine of the sort made by molecular manufacturing.

Nanomanufacturing: Same as molecular manufacturing.

Nanosurgery: A generic term including molecular repair and cell surgery.

Nanotechnology: See **Molecular nanotechnology**.

Positional synthesis: Control of chemical reactions by precisely positioning the reactive molecules; the basic principle of assemblers.

Protein design, protein engineering: The design and construction of new proteins; an enabling technology for nanotechnology.

Proximal probes: A family of devices capable of fine positional control and sensing, including scanning tunneling and atomic force microscopes; an enabling technology for nanotechnology.

Replicator: A system able to build copies of itself when provided with raw materials and energy.

Ribosome: A naturally occurring molecular machine that manufactures proteins according to instructions derived from the cell's genes.

Scanning tunneling microscope (STM): An instrument able to image conducting surfaces to atomic accuracy; has been used to pin molecules to a surface.

Sealed assembler lab: A general-purpose assembler system in a container permitting only energy and information to be exchanged with the environment.

Smart materials and products: Here, materials and products capable of relatively complex behavior due to the incorporation of nanocomputers and nanomachines. Also used for products having some ability to respond to the environment.

Star Trek scenario: Someone builds potentially dangerous self-replicating devices that spread disastrously.

Virtual reality system: A combination of computer and interface devices (goggles, gloves, etc.) that present a user with the illusion of being in a three dimensional world of computer-generated objects.

Virus: A parasite (consisting primarily of genetic material) that invades cells and takes over their molecular machinery in order to copy itself.

Acknowledgments

Many professors and students at MIT and Stanford University have helped refine the technical ideas underlying this book. Thanks also go to technical audiences at laboratories in the United States, Japan, and Switzerland for their critiques of these concepts.

A number of friends read early drafts and made suggestions for improving the accessibility of the book: James C. Bennett, Linda Boyd, Stewart Brand, Heidi Christensen, Allan Drexler, Helene Brun, Roger Duncan, Stan and Kiyomi Hutchings, Peggy Jackson, Jackie Kubal, Ralph Merkle, Russ Mills, Janice Morningstar, Ed Niehaus, Harold and Joy Pergamit, Amy Peterson, Gordon Peterson, Norma Peterson, Tracy Schmidt, Sue Schumaker, Carol Shaw, Leif Smith, and Marc Stiegler. To Barry Silverstein go special thanks for support during the writing of this book.

Thanks also to our transcriptionist Cheri Fjermedal, artist Pauline Phung, agent John Brockman, and editor Doug Stumpf.

Index

Page numbers in *italics* refer to illustrations.

Abelson, Philip, 88
abuse, 254–263
accidents:
    extraordinary, 251–254
    ordinary, 247–251
acid rain, 183, 279
activism, 279–280
Adamson, Sandra Lee, 220
AFM, *see* atomic force microscope
aging, 201, 224
agriculture, 171, 174–176, 188–189, 193, 196,
    239
AIDS, 215, 220, 221, 222, 279
Alcor Life Extension Foundation, 288
angstrom, 85
animals, 192, 231
antibody fragments, 118
Aono Atomcraft Project, 112, 113
archaeology, 145
arms control, 255, 256, 259, 262, 270, 278
Arnett, Patrick, 85, 95, 289
art, 238, 240–241
arthritis, 202, 213
artificial intelligence, 231, 287
assemblers, molecular, 62, 293
    chemical reactions by, 53

defined, 33–34
general-purpose, 33–34, 63, 140–141
limited, 257
natural analog, 100
primitive, 123–126
sealed lab, 258, 295
special-purpose, 69, 133
speed of, 124, 151
assembly, molecular, 61, 63
AT&T Bell Laboratories, 89, 95, 148
atherosclerosis, 212
atomic force microscope (AFM), 91, 92, 93–
    94, 109, 118, *119*, 293
atoms, 50, 75
Autodesk, Inc., 70, 107–108, 116, 148
automated engineering, 223–224, 293
automobiles, 149, 178–179, 236, 245

Babbage, Charles, 87
Becker, R. S., 95
Beller, W., 82
Benedict, Ruth, 237
Bennett, James C., 255–256, 287
Bergaust, E., 82
Biegelsen, David, 99, 123
Binning, Gerd, 93, 94
BioArchive Project, 225
biodegradability, 74, 143, 185

biomolecular science, 91
biostasis, 220, 224, 251, 288
biotechnology, 19–20
Bismarck, Otto von, 243
Blundell, Tom, 101
bonds, 53, 64
brain tumors, 219
bulk technology, 70
Burnham, John C., 36

cancer, 26–27, 39, 202, 204, 210, 279
Capasso, Federico, 148
capital equipment, costs of, 164
carbon, 73–74, 73, 150, 166–167, 191
carbon dioxide (CO_2), 30, 175, 182, 184,
    188–189, 191, 192
Carothers, Wallace, 102
catalysts, 103, 123
cell herding, 211
cell pharmacology, 293
cells, human, 55, 200–201, 202, 209–216
cell surgery, 214, 215, 216, 293
Center for Constitutional Issues in Technol-
    ogy, 255
Center for Genetic Resources and Heritage,
    225
change, 231–235, 242–245
chemistry, 71, 90–91, 104–107, 143–144
    role in nanotechnology, 104
chlorofluorocarbons (CFCs), 184
climate, 30, 175, 188
cloth, 159
common cold, 214
communications, 176–177
competition, international, 263
complexity, 80, 152, 223
computer-aided design (CAD), 70, 107–108,
    109, 113, 124
computers:
    costs reduced for, 29, 162–163, 242
    entrepreneurship and, 236
    mechanical, 87
    miniaturization and, 29, 148–149
    molecular manufacturing vs., 24
    molecular modeling with, 28–29, 44–56,
        91
    nanocomputers, 53–56, 131–133, 144–145
    technological development of, 85, 153, 244–
        245, 271
    see also nanocomputers
conference, nanotechnology, 78, 95
conveyor, 61
cost factors, 161–169
    for energy, 22, 25, 163, 165, 167–168,
        183, 188

for equipment, 138, 143, 164, 165
of greenhouse agriculture, 175
interdependence of, 166–167
of labor, 143, 162, 164
of molecular manufacturing, 161–169
per-unit vs. per-kind, 163
for product development, 151, 152
of raw materials, 163–164, 165, 166–167
technological predictions and, 40–41
of waste disposal, 164, 165
see also molecular manufacturing
Cram, Donald, 107

dangers, see accidents, abuse
dative-bond formation, 69
decentralization, 235–237
DeGrado, Bill, 89, 100, 101, 102, 103, 111,
    205, 289
Democritus, 75
dental care, 207, 213
Desert Rose Industries scenario, 128–138, 155,
    163, 164, 253, 254
design, costs of, 165
diabetes, 203, 217
diamond, 73, 73, 74, 150, 191
Diels-Alder reactions, 69
Digital Instruments, Inc., 93
disassemblers, 129, 294
DNA, 106, 195, 214, 223
    reader, 122, 215
Drexler, K. Eric, 77, 287, 288, 289–290, 291
drive shafts, 69
Du Pont, 87, 89, 100, 101, 102, 103–104,
    108, 204
dyes, tunable, 158

Ecological Engineering Associates, 185
economy, 231–242
    abundance, 237
ecosystem, 185, 194, 260
    protectors, 194–197, 259, 294
Edmondson, Daryl, 225
Eigler, Donald, 96
electrical power, 157–158, 160, 161, 163,
    166, 167–168
employment patterns, 238–242
endangered species, 279
energy costs, 13, 22, 25, 165–168, 183, 188
energy sources, 60, 157–158, 160–161, 166–
    168, 191
    see also solar energy
engineering:
    automated, 223–224
    compared to science, 110
    cut-and-try, 42, 43

engineering (*continued*)
  exploratory, 41–43, *42*
  genetic, 100, 193–194
  textbook, 42–43, *42*
  see also molecular engineering; protein engineering
Engines of Creation (Drexler), 6, 78, 94, 257, 258, 287
environment, 181–198
  agricultural land use and, 174, 175–176, 188–189, 193
  Eastern European damage to, 269
  restoration efforts for, 143, 181, 185–190, *187*, 193–198
  technological choices and, 260
  tunnel transportation systems and, 177–178
  see also pollution
Environmental Research Lab, 175
enzyme, 103
ERATO, 112, 291
Erickson, Bruce, 101
error rates, 64
ethics, 260
Europe:
  Eastern, 235, 236, 269
  economic alliance of, 236–237
exploratory engineering, 41–43, *42*, 54, 294
Exploratory Research for Advanced Technology (ERATO), 112–113, 291

Fahy, Greg, 215, 220
Fersht, Alan, 101
Feynman, Richard, 76, 290
First Foresight Conference on Nanotechnology, 95, 253
food, *see* agriculture
Food and Drug Administration (FDA), 263
Forbes, 241
forces, atomic scale, 49
forecasting, mistakes in, 39–41
Foresight Institute, 7, 78, 95, 225, 253, 285, 286, 291, 292
forests, 193
  destruction of, in U.S., 174
  extraterrestrial, 22
  restoration of, 30, 176, 188–189
  tropical, 29, 174
fossil fuels, 183–184, 188, 191
Foster, John, 85, 95, 96, 97–98, 289
fraud, 229
Frommer, Jane, 95
Frontier Materials Research Program, 112
furniture, 149, 159–160
fusion, difficulty of, 52
future, 21–22

difficulty of discussing, 37–41
unpredictability of, 38
Future Shock (Toffler), 243–244
"futuristic," 37

gases, 50, 72
gear, 53, 69
genetic engineering, 100, 193–194
genetic information, 31–32, 224–225, 227–228
Gerber, Ch., 94
germ warfare, 256
glass, 73
Global Business Network, 7, 78
global warming, 182, 188, 279
Golovchenko, J. A., 95
graphite, 73, 150, 191
gravity, 51
gray goo, *see* "Star Trek scenario"
Green Consumer, The (Elkington, Hailes, and Makower), 192
greenhouse agriculture, 174–176, 196
greenhouse effect, 184, 188
green manufacturing, 182
green wealth, 182, 183, 193, 198

Hahn, Karl, 103
harmonic drives, 69
heart disease, 212–213
hemophilia, 201
HIV virus, 221
Hofstadter's Law, 114
Horn, Paul M., 95
Hotani Molecular Dynamic Assembly Project, 112
housing, 172–174, 279
How Superstition Won and Science Lost (Burnham), 36
human rights, 230
hunger, world, 174
Hutterites, 230

IBM, 85, 93, 94, 95–98, *97*, 103, 104
immortality, unavailability of, 224
immune machines, 206, 223, 259, 294
immune system:
  nanotechnology applied to, 27, 205, 207–209, *208*, 210, 223
  natural functions of, 26, 27, 118, 202, 207, 209, 210, 213, 214, 222
industrial revolution, 159, 172, 233–234, 235, 239
industry, 239
  see also manufacturing
influenza, 214, 220, 222
information, 227, 240, 241

Institute for Physical and Chemical Research (RIKEN), 112
instrumentation, 88
insurance, 165
interfacial free-radical chain reactions, 69
international relations, scenarios, 32, 266–278
isolation policies, 233

Japan:
lack of term "futuristic," 37–38
magnetic trains in, 178
MITI, 17, 33, 86
nanotechnology research in, 17, 33, 111–113
protein engineering in, 101

Klis, Wieslaw, 103
Kobayashi, Prof., 17
Kunitake Molecular Architecture Project, 112
ku so no, 38
Kwakiutls, 237–238
Kyoto University, 112

labor, 143, 162, 164, 165, 239, 282
see also employment patterns
land use, 174, 175, 193
Leatherbarrow, Robin, 101
Lederberg, Joshua, 222
Lehn, Jean-Marie, 107
leisure, 237
Leonardo da Vinci, 87
life extension, 279
life span, human, 238, 242–243
limited assemblers, 257
limits, 227
on attention, 247
changing, 152
liquids, 50, 72
Liss, Alan, 185–186, 195
Lovelock, James, 182
Luddites, 233, 234
lupus, 202

machining, crudeness of today's, 70–71
McKenna, Terence, 182, 198
manufacturability vs. understandability, 43
manufacturing, 127, 239
see also molecular manufacturing
Marburg virus, 221
Massachusetts Institute of Technology (MIT), 78
materials, 144, 149–151, 191
see also smart materials
Mattick, John, 225

mechanochemical processing, 68
medical nanotechnology, 205–225
aging process and, 224
as cancer treatment, 26–27, 39, 202, 204, 210
diagnosis procedures in, 26, 203, 205
ethical issues in, 260
healing capabilities of, 21, 216–220
on human cells, 213–216, 223–224
limitations of, 219
metabolic regulation with, 216–217
nonoinstruments used in, 144, 145–146
outside tissues, 206–209
plagues and, 220–223
safety issues in, 250–251
software for advanced, 223–224
species restoration and, 224–225
surgery, 145, 203, 205, 211, 214–216, 219
viruses and, 26, 204, 214–215, 221–223
within tissues, 209–213
medicine, contemporary, 203–205
Megamistakes (Schnaars), 39
mental retardation, 219
Merkle, Ralph, 141, 253, 257–258
Merrifield, Bruce, 103, 106
metals, 70–71, 72–73
micromachines, 89
microtechnology, 71, 89–90
Milbrath, Lester, 182, 237, 254–255, 288
military technology, 32, 255, 256, 259, 262, 269, 270, 274–277
Miller, David, 87
Miller, Mark S., 291
miniaturization, 76, 85, 148–149
bottom-up, 90
top-down, 76, 89, 90
mining, 191
Ministry of International Trade and Industry (MITI), 17, 33, 86
mirai no, 38
MIT (Massachusetts Institute of Technology), 78
MIT Nanotechnology Study Group, 78
molecular building blocks, 99
molecular design, 108
molecular electronics, 122, 131, 149, 294
molecular engineering:
in biology, 91, 98–104, 204
building-block construction used in, 106
in chemistry, 104–107
computer-aided design used for, 107–109, 113, 124
research progress in, 111–116
tools developed for, 90–91
molecular herding, 97

molecular machines:
  in the body, 200
  computer models of, 113
  defined, 19, 76
  in nature, 19, 24, 98–100, 157, 171, 199–204
  origin of term, 76
molecular manipulators, 94–98, 112, 118–123, *119*, 124, 294
  drawback of, 121, 123
  speed of, 121
molecular manufacturing, 294
  abilities and constraints, 160
  in building computers, 148
  cleanliness of, 143, 182–188
  comparison with chemistry, 72
  costs of, 40–41, 138, 143, 151–152, 161–169
  defined, 20–21
  early, 59, 61
  efficiency of, 69
  energy for, 60
  error rates in, 64
  as example of exploratory engineering, 43
  precision of, 63–64, 70–72, 77, 152
  redundancy in, 69
  reliability, 64, 69
  scenario, 57–69
  speed of, 60, 66
  as terminology, 90
  see also Desert Rose Industries scenario
molecular medicine, 145
molecular modeling, 107, 109, 113
molecular nanotechnology, 9, 90, 294
  see also nanotechnology
molecular systems engineering, 110
molecular world, 56, 61, 65
molecules, defined, 33
muscular dystrophy, 194

Nagayama Protein Array Project, 112–133
nano-, 34
nanocomputer, 53–56, 131–133, 144–145, 148, 295
  electronic, 69
  inefficiency of mechanical, 65
  manufacturing of, 66
  simulation of mechanical, 53
  size of, 48, 51, 54
  speed of mechanical, 54
Nanofabricators, Inc., 58–60, 124
nanomachines vs. micromachines, 89
nanotechnology:
  abuse, 254–263
  compared to chemistry, 77
  compared to digital technology, 34
  compared to mechanical engineering, 77
  compared to space computers, 152
  conceptual origins, 77–78
  confusion about, 254
  costs, 138, 161–169
  criteria for judging, 37
  defined, 19–20, 34, 71, 90
  difference from microtechnology, 89
  economics of, 40, 138, 161–169
  environmental applications, 181–198
  ethics issues, 260
  as example of exploratory engineering, 43
  first conference, 7, 78, 95, 253, 291
  first course, 78
  in Japan, 17, 111–113
  lack of optimism on, 259
  likelihood of, 85
  limits, 227
  medical applications, 199–224
  motivation for discussing, 35
  nontechnical objections, 81
  obviousness of, 37, 41
  problems caused by, 231–247
  public policy, 278
  public understanding of, 35–37, 281–285
  regulation of, 256
  safety, 247–254
  social responsibility and, 264
  suppression as infeasible, 261
  technical book, 78
  technical objections, 78–81
  timing of, 24, 82, 84–86, 113–116
  use in medicine, 145
  use in science, 143
  weapons, 261
National Aeronautics and Space Administration (NASA), 138
National Audubon Society, 197
National Science Foundation, 44, 86
National Space Society, 220
Nature, 95, 96, 118
Nobel Prizes, 87–88, 93, 103, 106
nuclear power, costs of, 167–168
nuclear waste, 189–190
nucleus (atom's), 51, 52
nylon, invention of, 102

obesity, 217–218
oil spills, 183, 249–250
optimism, 36, 82, 171, 259
osteoporosis, 201, 213, 216
ozone depletion, 184, 186–188

parallelism, 98–99, 104, 123

parasitic diseases, 201, 210
Patterns of Culture (Benedict), 237
Pearson, Mark, 87, 103, 204–205
Pedersen, Charles J., 107
PERI (Protein Engineering Research Institute), 101
Pethica, J. B., 95–96
philosophy, 238
physics, 91, 144–145
Physics Today, 94
plagues, 220–223
pocket library, 58–59, 60, 125
pollution, 184
  assumptions about, 21, 22
  atmospheric, 30, 186–189
  from chemical industry, 250
  economic issues and, 182–183
  from heavy-metal elements, 129, 184
  from manufacturing process, 143
  from oil spills, 183, 249–250
  soil, 27–28, 129, 186, *187*
  water, 174, 186
Ponder, Jay, 108–109, 110, 111, 124, 290
Pool, Robert, 36
population, 171, 174, 191, 192, 229–231
Porter, Sanford, 197
positional synthesis, 105
potlatches, 237–238
poverty, 170–172, 279
  see also Third World
power generation, solar, 157–158
precision, molecular, 64
price elasticity, 241
problems, from nanotechnology, 231–247
products, 142, 143, 192
protein engineering, 99, 101–104
Protein Engineering Research Institute (PERI), 101
proteins, 51, 52, 99, 100, 101–104
protoassembler, 118–123
proximal probe, 91, 94–98, 113, 295
public policy, 284

quantum uncertainty, 79
Quate, Calvin, 94
Queensland, University of, 225

radioactive waste, 189–190
raw materials, 163–164, 165, 166–167
recycling, 136, 137, 184, 185, 191–192
redistribution, 230
regulation, 165, 255, 256–264
reliability, 64
religion, 282
replication, 253, 257

replicators, 138–140, *139*, 251–254, 257–258, 295
reproduction, rights of, 229–231
research, safe, 258
resource depletion, 21, 22, 28, 184, 191–192
ribosomes, 99–100, 295
Richards, Frederic, 124, 290
Richardson, Jane and Dave, 101
RIKEN (Institute for Physical and Chemical Research), 112
robots, molecular, 64, 65
Rohrer, Heinrich, 93

safety, 165–166, 247–254
Salin, Phillip K., 167
Sasabe, Hiroyuki, 112
Satellite! (Bergaust and Beller), 82
scanning tunneling microscope (STM), 91–93, 92, 94–98, 99, 109, 113, 271, 295
scenarios, 11, 25
  arms race, 32
  atmospheric cleanup, 30
  cancer treatment, 26, 39
  Chicken Little, 269
  coherence, 275
  international rivalry, 272
  molecular manipulator, 118–123
  molecular manufacturing, 57–69, 128–137, 253
  molecular world, 45–46
  ordinary expectations, 266
  Pollyanna triumphant, 268
  retirement, 243
  smart materials, 155–157
  solar energy, 25
  spaceflight, 30
  species restoration, 31
  "Star Trek," 252–254
  supercomputer, 28
  Third World, 29
  as tools, 25
  toxics cleanup, 27
  uses of, 7
Schnaars, Steven, 39
science:
  engineering vs., 106, 108–110
  nanotechnology applications in, 143–145
  popularization of, 36–37
Science (journal), 88
scientific visualization, 44–45
sealed assembler lab, 258
self-assembly, 99, 103, 107, 113, 124
services, 239
shorai-teki, 38
Silicon Valley Faire, scenario, 57–70, 80

skin conditions, 206, 212
smart materials, 75, 154–161, 186, 295
social responsibility, 264
Social Security Administration, 243, 282
software, 110, 126, 177, 244
    for advanced nanomedicine, 223–224
    for assemblers, 130, 133
    for computer-aided design, 107, 124
    cost of development, 236
    DNA as genetic, 225
    limitations on, 152
    molecular modeling, 91, 108
    for nanomedicine, 211
    speed of development, 151, 244
    U.S. expertise in, 274
    see also artificial intelligence; automated en-
        gineering; complexity; computer-aided
        design
solar cells:
    cost of, 22, 25, 163, 166, 183, 188
    for environmental cleanup, 30, 187, 188–
        189, 250
    on road surfaces, 25, 30, 136–137, 157–
        158, 166, 183, 188–189
solar energy, 157, 158, 183
solids, 72–75
space, 138, 179, 189, 279, 287
spaceflight technology:
    cost of, 22, 30–31, 40, 179–180
    progress in, 151, 152
    structural materials for, 149–151
species, problem of imported, 193–198
species restoration, 30–31, 224–225
spinal injuries, 201, 218
Stanford University, 78, 94, 138, 139
"Star Trek scenario," avoiding, 252–254
Stewart, John, 103
STM, see scanning tunneling microscope
strategies, 7, 84–128, 261, 279–285
surveillance, problem of, 259
Swartzentruber, B. S., 95
Swedish Academy, 93

teamwork, 109
technological fix, 6, 22
technology, "high," 182
Temin, Howard, 222
terrorism, 256
thermal vibration, 50, 62, 64, 79
Third World, 21, 29, 162, 171–172, 173,
    210, 230
Toffler, Alvin, 243–244
Tokyo, University of, 106
Tokyo Institute of Technology, 112
toxic waste, 184, 279
    see also pollution
trade, 237
transistors, 131, 132
transportation, 136, 177–180
Tullock, Gordon, 40
tunnels, 136, 177–178

viruses, 26, 204, 214–215, 221–223
virtual reality, 29, 45, 295
    molecular world, 45–56

Walker, John, 70, 107, 114, 116, 148
Ward, Michael, 108
warfare, 256
    see also abuse
waste, 164, 165, 182, 183, 184, 189–190,
    191–192
water, 50, 174, 175, 186
wealth, 181, 237, 243, 247
    see also green wealth
weapons, 256, 259, 261
    see also abuse
Weinberg, Alvin, 168
Whole Earth Review, 182, 198

Xerox Palo Alto Research Center, 99, 123,
    141, 253

Yale University, 108, 124
Yoshida Nanomechanism Project, 112